AWAKEN

THE

GENIUS

Mind Technology for the 21st Century

Patrick Kelly Porter, Ph.D.
Foreword by
Brad Steiger

This book is dedicated to every person striving to make a difference in the world. To geniuses awake and asleep--may we all share in the the delight of discovery.

Awaken the Genius Foundation
c\o 309 Aragona Blvd., Ste. 102-712
Virginia Beach, Virginia state PZ[23462]
757-499-5097 Positive Changes / 800-880-0436

Other Products
By Patrick Porter, Ph.D

Psycho-Linguistics (Book)
The Power of Your Voice (Book)
Adventures in Accelerated Learning (Tapes)
Adventures in Self-Discovery (Tapes)
Personal Evolution Series (Tapes)
Awaken the Genius Audio Book

For complete information about
Awaken the Genius Seminars
or for a catalog of self-help tapes, write:

Awaken the Genius Foundation
c\o 309 Aragona Blvd. 102-712
Virginia Beach, Virginia state
Postal Zone/Code [23462]

Illustrations: Dean Sylvia
Cover Art: Sam Johnson

Disclaimer: This book is designed to provide information in regard to the subject matter covered. It is sold with the understanding that the publisher and author are not engaged in rendering psychological advice and the processes in this book are non-diagnostic and non-psychological. If psychological or other expert assistance is required, the services of a competent professional should be sought. The purpose of this book is to educate and entertain. Neither Awaken the Genius Foundation, the author, or any dealer or distributor shall be liable to the purchaser or any other person or entity with respect to any liability, loss, or damage caused or alleged to be caused directly or indirectly by this book. **If you do not wish to be bound by the above, you may return this book for a full refund.**

ISBN: 0-9637611-8-8
Library of Congress Card Number: 93-086044
Printed in the united States of America

10 9 8 7 6

TABLE OF CONTENTS

CHAPTER NINE

CHAPTER TEN

CHAPTER ELEVEN

CHAPTER FIFTEEN

ACKNOWLEDGMENTS

Genius is a higher order of thinking. Beyond survival and success there is a synergy that we all draw upon. This work didn't start with me and most certainly won't end with this text. Surely Einstein and Tesla and the other great adventurers didn't invent it either--it is an inherent part of human nature.

I am grateful to have been born into such a phenomenal time known as the "information explosion," and to share in the techniques that helped me to awaken my learning potential. I am grateful for the inspiration and wisdom of many thinkers and trainers who have shaped my life. I would like to personally acknowledge all my teachers, some of whom I have met, and some of whom I have not met (yet). The most important include Swami Sri Sathya Sai Baba, Paramahansa Yoganada, Hamid Bey, Paul Adams, Gil Gilly, Tony Moltzan, Patricia-Rochelle Diegel, J. J. Hurtak, Milton Erickson, Richard Bandler, John Grinder, Steve Andreas, Robert Dilts, Tony Robbins, and Jerry DeShazo. I am grateful for the many students, friends and clients who have continuously proven the usefulness of this material.

I am grateful to have been born into a family with eight brothers and sisters. Michael, Michelle, David, Walter, Sarah, Bill, John and Fran -- all truly awakened geniuses. I honestly feel blessed that my father was an alcoholic and was thus motivated to get into the field of mental wellness. Through his example he freed nine children to live proactive lifestyles. I am grateful to my mother, who spent the greater part of her short life taking care of such a tribe. She was a fearless leader.

I am grateful to my wife Cynthia, who spent endless hours helping to put together the *Awaken the Genius* seminar and book. Without her special creativity we would all miss out. Thank you Cheree and Alex, my own two geniuses in training, for helping to explore the usefulness of these techniques.

A special thanks to my editors, Dan Nelson, Christa Nelson, Erik Nelson, Cynthia Fertal, Jon-Terrance Diegel and Professor Alice Adams of Prestonsburg College. Thanks for your special notes and comments. I humbly thank Brad Steiger, a great author and philosopher, for his thoughtful foreword to this book. I also offer my gratitude to Vivien Howe for the opportunity to create the *Awaken the Genius* seminar, to Dean Sylvia for sharing his unique artisitic talent and to Gerry McLarty for allowing me to delve into the mind of a genius. I would also like to make special thanks to Rita Livingston and Tyler Clements--two fantastic networkers.

Special thanks to Dr. John Sanders of Pikeville College for his time, talent and effort in researching and proving the effectiveness of these ideas and techniques in an academic setting.

I am grateful to you, the reader, for creating the motivation to finish this book. I confidently await your contribution to this remarkable planet as you awaken your genius.

-- Patrick Porter

FOREWORD

Since the societal stimulus of the late 1960's with its colorful time of protesting "what is" and seeking to discover "what might be," increasing numbers of people have been directed by an inner urging to find better ways of restructuring their lives toward self-realization and self-mastery. This deep desire to achieve a more complete mind-body-spirit balance has survived numerous cultural fads and fashions and proliferating numbers of self-styled gurus who promise simple, smooth, and safe pathways to instant awareness.

Today it often seems as though the demand for better things and for better living has multiplied the number of these gurus to the point where there appear to be more teachers than students. Stay up a little late some evening and spend an hour or so channel surfing the many infomercials that promise you easy-to-achieve methods of gaining wealth, opportunity, and enlightenment--all for the simple exchange of your credit card number.

All the while you are listening to these coaxial cable counselors tell how effortlessly you may attain your goals, and are watching their self-satisfied smiles show you how happy and comfortable they are, you will probably not hear one of them advise you that personal success and self-improvement--to say nothing of self-mastery--require commitment, discipline, discretion, and devotion.

Anyone who no longer lives at home with Mom and Dad understands that you do not get something for nothing in this materialistic world of ours. Even regular earthly existence requires some sort of barter, labor or money for goods and food. Surely the higher goals of the secrets of better living or higher awareness of necessity must require some tangible (i.e. monetary) exchange as well. But with such a seemingly infinite number of available guides, gurus, and instructional manuals from which to choose, how do you go about selecting a master, method, or course of study that will lead you to higher levels of awareness, rather than deeper levels of financial indebtedness?

Make no mistake about it, you are going to have to buy this book that you now hold in your hands--if you have not already done so. But, let me assure you that you will find it to be an extremely useful text and an excellent investment for your dollars spent.

In ***Awaken the Genius***, Patrick Porter does not claim to be a master from some lofty plane of existence who has all the answers available for his personal arbitrary dispersal; nor does he promise that self-mastery may be achieved by the repetitious chanting of magical affirmations of twenty-five words or less. But what he does offer are valuable tools which you may use to work steadily toward the achievement of your own self-mastery.

Dr. Porter assures us that we each have a "genius" within us. However, he recognizes that my genius may access information and utilize it in a manner somewhat different from yours. That is to say, some of us process sensory input more effectively through visual means, others through auditory stimuli, and so forth.

Through extensive--but not at all formidable--exercises, techniques, questionnaires, sample dialogues and lots and lots of anecdotal illustrations, Dr. Porter leads us step-by-step to a greater understanding of ourselves and *how* and *why* we may have created our own stumbling blocks and limitations. Then, thankfully, he demonstrates how we may free ourselves from the muck and mire of our self-created psychic mud fields.

By drawing from his own experiences, Dr. Porter is not at all reluctant to demonstrate to the most discouraged or depressed reader that in the past he, too, has been mired in his own self-made mud ...that he, too, has had to work to achieve a better way ...that he, too, had to practice discipline and discernment to awaken his own genius within.

If you are seeking a method of self-improvement that will immediately address your present self-imposed limitations and compassionately invite you to step into the "Circle of Power" that can allow you to achieve your loftiest goals, then Patrick Porter's *Awaken the Genius* may well be your doorway to the kind of tomorrow that you have always dreamed might be yours.

Brad Steiger

INITIATION

THE PHILOSOPHY OF A GENIUS

It was what Christmas morning should be. The brilliant sunshine reflected like diamond dust on a blanket of new fallen snow. Mrs. Johnson rolled over and faced the window. She squeezed her eyes tight, not ready to face the bright morning. Then, as thoughts of her husband's "plan" flooded her mind, she blinked her eyes wide open. All of the morning drowsiness was gone. Susan Johnson was very much concerned about how her sons would react to her husband's very different Christmas proposal. She had not fully agreed to any such test but he insisted upon proving his point. Susan was no longer sure just what Jack's point was. Sleep had been elusive that peaceful Christmas Eve. Dreams filled with fear and worry danced round her mind all through the night. What kind of damage might Bob endure through this peculiar game devised by his father?

Jack Johnson rolled over and gazed past his wife's shoulder. "It's a perfect day," he said, smiling. "Let's see if the boys are up."

As Jack slowly pulled the razor across the stubble on his face, he thought over his plan. The boys' gifts had been carefully wrapped by Susan the night before. Jack's mind was now on the contents of the boxes. For Zach, who seemed to have been born a pessimist, every gift he had asked for was wrapped beautifully and placed carefully under the tree. As for Bob, the Johnson's optimistic, and fun-loving younger son, Susan had wrapped a small, square box in ordinary red paper with no ribbon. A grin crept across Jack's face as he imagined his son's face upon opening the gift.

Susan had argued against the plan for days. "It's just bizarre!" she had scolded her husband. "No child deserves that kind of treatment on Christmas morning!"

Strange as it was, Jack Johnson convinced his wife that everything would turn out all right. Susan finally agreed after Jack promised that Bob would receive his real Christmas gifts later that day.

When it was time to open the festive holiday packages, and everyone had gathered around the tree, Susan's eyes began to moisten. She gazed at Bob through wondering eyes. Bob sat with his back straight and his broad smile reflected the inner excitement he was feeling. He knew that something magical was about to happen.

Zach, on the other hand, was busy digging through the gifts, searching for those with his own name on the tag. He was mumbling to himself, "It'll probably be just like every other year ...a bunch of junk I'll never use."

Jack Johnson felt a brief flicker of uncertainty. "Bob," he said, in his most fatherly voice, "You open yours first."

Susan glared at her husband, then turned toward Bob. Her lips tightened into thin white lines. She tried to prepare herself for the moment Bob would lift the top off the box. What will he think of us?

Bob's excitement grew as he unwrapped the box. Since there was only one gift for him, he knew it had to be something big. He yanked off the lid and gazed down into a small pile of horse dung.

Susan and Jack Johnson watched in shock as their son jumped up, deftly dodged the gifts that Zach had strewn around the room, grabbed his coat off the hook, and ran out of the house. Before the Johnson's could respond to Bob's odd behavior, Zach began ripping through his presents. He quickly tore open each gift, yanked the toy or clothing out of the box and promptly tossed it aside. With a look of disgust, he would move on to the next. It took him all of two minutes to rip every gift open, then he turned to his parents, "Is that all I get?" He whined.

This reaction was no surprise to either of the Johnsons. It was what they had come to expect from Zach.

But Bob's response they did not anticipate. Jack was no longer certain of his plan at all. Where could Bob have run off to, and why? This was not good at all. Susan just sat there, numb, staring at her husband. Her face changed from ashen to crimson before she rose from the chair and went to the hall closet. She clumsily pulled on her boots and coat and then went out the back door in search of her son. Susan never felt the icy sting of tears turning to ice on her cheeks.

By the time Jack had found his own warm boots and parka and then stepped out into the brilliant sunshine, his wife had already reached the dense woods at the edge of their yard. Susan had easily followed Bob's footprints to the edge of the thicket, but the trail came to an abrupt halt there. "Susan," Jack pleaded as he reach her side, "please talk to me."

"What's going on around here?" Zach called from his place in the living room. Up to that moment he had been oblivious to the other members of the family. Finally he sniffed, shrugged his shoulders and swiped his nose with the sleeve of his pajamas. He again removed his family from his mind and turned his attention to the abused pile of toys beside him.

The Johnsons looked everywhere for Bob, but he seemed to have vanished. Zach had finally grown curious enough to join his parents in the search, but still they found no sign. Then they remembered the barn at the edge of the wooded lot. It was a decrepit old shack that had once sheltered the farmer's horses. There they found Bob. He appeared to be searching every corner of the barn. His boots, jeans and gloves were smeared with aged horse dung.

"Bob," his father called, "What are you doing out here?"

Bob glanced up at his parents and brother. His eyes were gleaming with excitement and his mouth circled in his familiar grin, "I know there's got to be a horse here somewhere!" He called back.

***

Very few people ever reach the status of "genius." There are many factors that prevent people from using their full genius potential. I believe the.main deterrent is *fear*. Fear comes in many shapes, forms and sizes. Fear infiltrates, paralyzes and destroys; it oozes into the deepest core of the human potential, freezes all creativity, and then expands out into every thought and action. Then comes the most destructive fear of all -- it exists in the feeling of *lack*.

Feelings of lack can start at a very early age, especially in the "designer" world of the 90's. Children begin experiencing feelings of lack when they perceive other children's clothing, homes and automobiles as nicer than their own. Sometimes it starts with the belief that they are lacking in their own intellectual abilities. I believe that somewhere along the path of growing up we begin to confuse need with want. Like the pessimist, we will then develop negative thinking patterns about ourselves, the world and other people--especially those who have what we want. Zach may be an extreme example, but have you ever known young children with negative attitudes like his? These negative thought patterns, even to a small degree, are destructive and will block the genius potential.

Intelligence starts with the *"awe" response*, as was seen in Bob's reaction to his gift. It never dawned on Bob that he might have been "dumped on," as so many people might have felt. He assumes life is good. Bob looked upon his gift as a part of something wonderful, a horse. I guarantee you that Bob will grow up to be successful at whatever he chooses to do. Success will come to Bob because he won't stop until he has it. If Bob's positive attitude and optimism are encouraged, he may even be "destined" for greatness.

Awaken the Genius is dedicated to people who are ready to acknowledge that life is a series of problems and are willing to create solutions. This book was created for those who are ready for 21st century thinking; those who want to step out of the prison of their conscious mind and into the world of their imagination; those who know that it is possible to awaken the *other* level of their mind--the level that exists outside of time and knows no restrictions or limitations.

The status of "genius" is attained with the realization that personal growth and the attainment of wisdom are the main ingredients to a successful life. It is the growth and achievement that makes life an adventure. Most importantly, geniuses realize that *now* is the best starting point for a brighter and more exciting future.

I invite you to awaken your genius potential. It's there, within you; I know it is. We are all imbued with the ability to shine. Start by shaking your old beliefs loose, and focus on the kind of genius that exists within you.

If you find it hard to believe that you could possibly be a genius, think again. When I was in the second grade, I was held back and labeled "learning disabled." I was a "Zach." I was the class clown. I considered myself a "victim" of the system. What I have done with my own life is nothing short of a miracle, and I will share with you the many personal challenges and triumphs that created me; a person able to achieve two doctorate degrees; someone capable of writing this book; and someone who is fulfilling a life-long dream--a career helping others. Believe me, I didn't awaken my genius over night, but it did happen and is still happening.

The truth is, we are all moving; some people are going forward, some backward, and some wherever the advertisers tell them to go. Are you ready to realize the limitlessness of your potential? The only "limits" are those you choose through your own thoughts, actions and beliefs. A genius knows life is a game. There are no winners or losers, no victims or martyrs; life gives in the exact proportion that we ask. Awakened geniuses are simply people who have learned to ask the right questions--of the world and of themselves. Time is their ally.

My father once told me that a person who works from the neck down can expect to earn up to $20 per hour, and the person who works from the neck up really has no limits as to what can be earned. This made a lot of sense to me, especially after seeing my father come home from the factory day after day complaining that he wasn't making enough money. At that time he was truly stuck. He was actively living out what he perceived his life to be. He spent half of his life at the factory with people just as stuck as he was, and the other half in a bottle. He remained that way, trapped in a life of his own making, until years later when he became willing to trust in his own genius. He stopped drinking with the twelve-step program of Alcoholics Anonymous, went back to college, and then started a business helping people to help themselves.

When you trust in your mind it will be true to you. Every genius had to take that first step, and it came with the recognition that they are far more capable than they have been led to believe. They became willing to stretch the limits to the max in order to achieve their goals.

I once attended a seminar on prosperity. Truthfully, I don't remember much about the seminar, but there was one segment that made an impression on me. The presenter made a strong correlation between the concepts of *time* and *worth.* I had never thought of the two as being related. At the time I was still going to school, was working a full-time job as a warehouse supervisor, and was working part-time with my father whenever possible. I wanted to get into the therapy field full-time, but something stopped me from taking that final step of quitting my full-time job. Some might call this a block; others might say I was afraid. At the prosperity seminar, I came to realize that it was neither of these. I had simply found my *comfort level*, and was not yet ready to break outside of its safe boundaries. I was making relatively good money as a supervisor. I was single at the time, so it was easy for me to meet my few expenses. The cost of my education was covered, and I even had a car that almost always got me where I wanted to go. Indeed, it was difficult to step out of those comfortable boundaries to start a counseling career. Yet with this realization I knew, more than ever before, that I had to do it.

Monday morning, 6:00 a.m., I marched into my boss's office and handed him a letter of resignation. He gazed up at me with that superior, all-knowing look that only a boss can have.

"Aw, come on, Porter," he said, "Nobody quits a job like this!"

"I do," I said calmly, as I turned on my heel and walked away.

"Do you know how many other twenty-four year old kids there are who would give their eye tooth for this job?" He was talking to my back now.

"You'll be back!" I heard him shout, after his office door clicked shut.

I never went back--and my life has been moving forward ever since.

What was it that convinced me to step outside of my comfort level? It was the realization that when you break *time* down into its true *worth*, you are already prosperous! It's a simple matter of focusing your time and energy on what you want.

There have been many mentors in my life -- they were there because I knew that great geniuses understand the value of learning from others. One such teacher is Dr. Gil Gilly, and his classroom is the world. Whenever Dr. Gilly meets new people, he greets them with a firm handshake and then slips them a little card. The card gives a fantastic representation of time. I am certain Dr. Gilly's little cards have changed many lives. Take a moment and contemplate its meaning for you.

BANK OF TIME

If you had a bank that credited your account each morning with $86,400, that carried over no balance from day to day, allowed you to keep no cash in your account, and every evening canceled whatever part of the amount you failed to use during the day, what would you do? Draw out every cent, of course! Well, you have such a bank, and its name is "Time." Every morning it credits you with 86,400 seconds. Every night it rules off, as lost, whatever you have failed to invest to good purpose. It carries over no balance. It allows no overdraft. Each day it opens a new account with you. Each night it burns the records of the day. If you fail to use the day's deposits the loss is yours. There is no going back. There is no drawing against "tomorrow." You must live in the present, on today's deposit.

You have made an investment in yourself and your family by purchasing this book. It is more than the price paid; it is an investment in your **time**, which is one commodity you cannot buy. Time is given to you free of charge, and what you do with it is up to you. In the words of the great world teacher, Sri Sathya Sai Baba, *"Those who want to secure pearls from the sea have to dive deep to fetch them. It does not help them to dabble among the shallow waves near the shore and say that the sea has no pearls and all stories about them are false."*

What is meant by this simple example? That there will always be those people who believe in the struggle of life and little else. They will tell you, *"That's impossible!"* and *"You can't do that!"* or, *"Accept it, kid, that's just the way life is--you can't change it!"* To these people, there is only one way to learn--their way. These are the people who are not willing to dive deep to fetch their pearls.

I am here to tell you that the finest, shiniest pearls are the ones you must dive into the deepest depths to find. They are there, just waiting for you to fetch them. You can consider the chapters and techniques in this book your diving guides. These guides will take you on the most direct dive to your pearls (goals). They are simple and profound. Indeed, once put into action, I have never found them to fail, not even once in eleven years of helping people just like you. Life becomes fun--it's an adventure--and in this book you will learn how to find all of the pearls that are rightfully yours.

The secret is in putting it all together to work for you; to create a state of *infinite abundance.* This simple statement, infinite abundance, means that there is *plenty*--for you and for me, and plenty of everything. Unfortunately, infinite abundance is not the present belief of our society. We all possess the ability to attract into our lives what we need. However, this is not always what we want. This is how people get themselves into trouble. Probably the greatest tool the genius possesses is the understanding that we are always in the right place at the right time, and will always have what we need. This may be a difficult concept for those people who have always held true to their philosophy that someone or something else controls them.

Some people believe that their parents, spouse, friends, teachers, the government or other authorities are in control of their lives. They have a whole arsenal of excuses and are loaded and ready to blame another person for their shortcomings. They will tell you that their mother didn't love them or that their father was never home, that none of their teachers ever cared about them and most people, they are certain, were out to get them.

Still others allow their lives to be controlled by circumstance. These are the *"if only,"* people. I would have ...*if only* my parents had not divorced when I was ten. I could have ...*if only* it had not rained that day. I would try ...*if only* I had better luck. I could get a better job ...*if only* I lived in a different city. I'm sure my life would work .... *if only I could win the lottery!*

The habits of self-imposed limitation and of blaming others are probably the main reasons why there are so few people in the world recognized as "Genius." The reality, however, is that every one of us is encoded with the genius potential. This capacity is natural and readily available within us at our summoning. It is very real and even tangible once one becomes aware of its presence. This potential can be likened to a seed. Within one seed from an oak tree is the potential for millions of mighty trees. Yet, if it is not germinated, it can't possibly sprout, grow roots, build a trunk and branches, form leaves, or create the seeds for its future generations. I find that as I regularly dive for pearls and nurture the seeds I have planted, the opportunities I need are always there; often before I know that I need them!

After I finished my education, I was presented with the opportunity to work with the Arizona Health Council, a group of therapists who work with educational programs for the state. What the state needed was a program utilizing all of the new "proactive" processes that integrate the right and left hemisphere of the brain. **Awaken the Genius** was born first in seminar format through my work with the Arizona Health Council. In all honesty, we were amazed with the results. The processes were enjoyed not only by students but by people from all walks of life, including teachers, parents and business executives. My hope is that as you read through the text, you will achieve what hundreds of seminar participants have; you will activate that part of your mind that works outside of conscious awareness and put your "whole brain" into action.

While in the midst of writing this book, I experienced one of those magical moments when everything seems to connect perfectly. I had just bought a new encyclopedia on Compact Disc for my computer. As I was leisurely scanning through its contents, I "accidentally" made a most exciting discovery for my research on activating the genius. My mind was actually a million miles away from any thought of this book, but there was what I had been seeking, boldly leaping from the computer screen and into my awareness. What I found was not listed under "genius," it was in the section called "mythology"--a place I would never have considered to look for genius stuff. In a nutshell, this is what it said:

In Roman mythology a ***Genius*** *was a guardian spirit that protected an individual throughout his or her life. Every living person was endowed with a specific genius to whom yearly offerings were made, generally on the person's birthday. In addition to the individual's genius there were genii who protected tribes, towns, places, and the Roman state. A particularly important genius was the Genius Populi Romani, guardian of Rome. The accomplishments of an individual were often attributed to his or her genius.*[1]

[1]The New Grolier Multimedia Encyclopedia.

Although this Roman myth about geniuses being guardian spirits might not be all that believable today, it does, like most mythology, draw upon some deeper inherent truths. We all possess an inner guidance, one that is greater than the conscious mind. When you learn to tap into this power, you are, in all reality, awakening your genius. You will know it when it happens, and others will consider you a genius as well. This part of your mind is beyond what is conscious to you and is aware of even the most minute details of your life. There is only one thing stopping you from accessing this genius part of the mind; you were never taught how to access and use it.

Think of a computer for a moment. Even with all of its incredible capabilities, it can be instantly rendered useless if you remove all the programs. The full potential is still there, inside the computer, but there is no software to make it work. Is this not how we are all born? We are equipped with the hardware, but with very little knowledge of the software. Most of the programs were built by our parents, family and friends. These programs may or may not be working in their lives, but if they're not, how on earth can you expect them to work for you?

As you read this book you will discover that it won't matter whether your parents are scholars, factory workers or alcoholics, the applications found here will help you to understand the mind and its awesome, never-ending ability to work for you. You may find yourself setting goals and achieving them, without ever knowing you have put forth the effort. Whether you want to become an academic wizard, build a perfect memory and recall, simply learn the processes of time management, or become President of the United States, the basics are all here. Enjoy, and allow yourself to experience your mind to the Ph.D level. You can only attain the "Ph.D of the Mind" from applying the techniques for yourself.

Good Luck, Genius. You have already made the first step.

Love & Light,

Patrick

QUESTIONS AND ANSWERS

Question: What is the best way to use this book?
Answer: *The best way is different for everyone. Let me give you an example: There are two ways to eat an elephant. You can swallow it whole and allow it to slowly digest, perhaps with a great deal of discomfort. Or you could cut the elephant into small pieces, chew it completely, allow for proper digestion and assimilation. If you are the type of person who picks up every new book written about the mind, mills through it very quickly and promptly forgets what you have read, then I challenge you to read through this book slowly. Take one section at a time, assimilate what you've read, do each of the Self Discoveries and take time to listen to each Self-Help Dialogue. I assure you, investing this time will come back to you one hundred fold.*

But why take my word for it. Take just a few moments for the ***Self-Discovery*** *following this section. You will convince yourself.*

Question: What is a Self-Help dialogue?
Answer: *The Self-Help dialogues are designed to be spoken by you into a tape recorder and then played back while you simply relax and listen. This will accelerate your ability to assimilate the information and will change any negative self-talk that might be preventing you from awakening your genius. These transcripts are written in the format of a third person dialogue. When you are listening to the tapes, your voice takes you on peaceful journeys filled with wonderful new discoveries and transformation. The sound of your own voice becomes familiar and comfortable to you, like hearing the caring guidance of a close personal friend.*

Question: What if I have a terrible voice?
Answer: *If you believe your voice is terrible, you have all the more reason to benefit from doing the exercises and creating your own self-help tapes. Believe it or not, your body and mind crave the sound of your own voice. Not a harsh commanding voice, but a voice filled with love, compassion and understanding. It's time we stop "shoulding" all over ourselves and start to use the mind the way it*

was intended. Your voice will work as a guide and your mind will listen to you more efficiently than it will to any other voice. You may be surprised at how, with practice, your voice can become pleasing. Good luck, and remember that the tapes are for your personal use. Have fun with them. Enjoy the time to yourself and, most importantly, relax.

Question: Do I have to do the **Self-Discoveries**?

Answer: *It's your life--you can do whatever you want. I have found that when a person has an experience, the learning is much more likely to be retained and the benefits will stay with that person throughout life. If you want a permanent, lasting change in the way you think and process information, dive deep, explore and risk. You have nothing to lose but perhaps a little ignorance.*

Question: What is a positive intention?

Answer: *"Positive intention" (as outlined in this text) is used to describe the core purpose for a behavior or belief. For example, the positive intention behind studying is to ensure that you know the information for a test. This positive intention is what motivates a behavior or action Other forms of positive intentions, on the other hand, can create unwanted behaviors. An underlying concern about failure might have procrastination as its outer behavior. In other words, if the person never gets around to trying, failure can never happen--but neither can success.*

It is important to realize that one hundred percent of the time we are doing the very best we can with the information we have at hand. The purpose of awakening your genius is to provide you with new information that will serve you and help you to become proactive instead of reactive to life's problems. Proactive means that you take control and create solutions, even out of the most difficult problems.

SELF-DISCOVERY: THE VALUE OF YOUR TIME

Take a moment to fill out the following questionnaire. Take your time and think your answers through.

How much is your time worth?

$______________ per hour

Now think of how many hours in the last month you spent procrastinating or sabotaging yourself. _____ Total. Convert that into dollars using your per hour rate shown above.

$______________ Total

If you were to take that money and invest it in yourself, how much would it be worth to you in one year?

$______________

How much in 10 years?

$______________

What are some of the things you can do in the future to utilize this valuable commodity called *TIME*?

***

I have heard so many people say, if only I had this skill or that knowledge, I might also experience success. I agree, life would certainly be much easier if simply obtaining an ability or gaining new knowledge would guarantee success. Unfortunately, this isn't necessarily true. Awakened geniuses know better than to measure achievement by an end result. The truly awakened genius has learned to literally embrace success through a series of day-to-day actions. These are actions that can be tested and repeated and that bring about a feeling of progress. Geniuses recognize that their time is of great value; they can't afford the luxury of wasting a single minute. As you actively awaken your genius, through the action of reading and assimilating this text, you can enjoy the on-going enlightening (lightening of your mental load) that comes naturally when you are making a difference.

"Do you want to know the great drama of my life?
It's that I have put my genius into my life;
all I've put into my works is my talent."

Oscar Wilde
1854-1900

CHAPTER ONE

WHAT IS A GENIUS?

Thomas Edison, known the world over as the inventor of the light bulb, was once asked how he managed to keep himself motivated while working on a project. Edison recalled that when inventing the light bulb, he had experienced 999 failures prior to his discovery. His response was a simple but profound one. "I didn't fail 999 times at creating the light bulb," he replied; "I found 999 ways in which it just won't work."

What did Edison have that seemingly eludes most people? His ability to restructure his thinking so that no matter what happened, it was a success; and what others would perceive as failure was, to him, *feedback*. It is this attitude that creates a genius.

Sound a little too simple to you? It can't possibly be *that* easy to become a genius, you might say. Perhaps you thought that to be a genius you must first pop the top off the IQ charts. Well, read on.

Did you know that Edison even went so far as to attempt creating a light bulb out of peanut butter? The pragmatists must have rolled their eyeballs and exclaimed, "What a waste of time!" True, these efforts failed miserably, but Mr. Edison was not *really* trying to invent a light bulb with peanut butter; he was freeing his creative mind to formulate a solution. He may not have understood the "method behind his madness," but psychologists now understand that he was balancing the power in his mind between the right and left hemisphere.

BALANCE OF POWER

Have you ever sat in a classroom listening to a lecturer and about half-way through noticed yourself growing so exhausted that you just had to let your mouth stretch into a yawn. Perhaps your head began to feel like a lead weight, as if your chin was magnetically drawn to your chest. This type of sleepiness is caused by the overload placed on the left hemisphere of the brain. The left brain can be likened to a computer filing system -- with a few vast differences. Human intelligence is far superior to even the most sophisticated computer system for one profound reason--the brain not only stores and processes information, it also records each bit of information with *emotion.* In this very real way your brain stores information holographically.

Pause here for a moment and think about "**apple pie**."

Perhaps you can see the apple pie in your mind or imagination. Maybe you are able to recall the sweet, apple-cinnamon scent of a pie just coming out of a hot oven. Were you, perchance, like one seminar participant who stood before the group and described how she could *see* her grandmother's broad smile, and her face, a multitude of crinkles that reflected many years of joy, worry, laughter and sorrow. She told us about Grandma's frail little arms that were tucked into huge oven mitts and of the *love* she *felt* as she watched Grandmother slowly work her way to the dinner table with her beautifully crafted creation. The family would *ooh* and *aah* appreciatively as they *inhaled* the rich aroma and allowed Grandma's steaming apple pie to *melt in their mouths.* This young woman was remembering hot apple pie holographically--with all the senses of sight, smell, taste, touch, and sound involved.

All of this holographic processing takes energy--mental energy. It would be logical then that a key ingredient to genius would be to find a way to think holographically, but without expending so much mental energy.

I feel fortunate to have met and become friends with a true genius. This gentleman is extremely gifted in the field of computers and in the art of invention. When I first met Jerry, I considered him rather eccentric. His home, inside and out, was stacked with junk. I asked him how he could find anything in such a mess; let alone come up with inventions from an absolute scrap heap. *(Jerry builds all of his prototypes from scraps and old parts.)* He related an extraordinary story.

When Jerry was a small child, he would fall into deep, peaceful sleep and dream of all that he wanted to build. This in itself was not uncommon, as most young boys at one time or another will dream of fame and riches as a renowned inventor. Unlike other boys, however, Jerry never grew out of it. With each passing year his dreams grew more vivid. This slumber world became real to him as he developed a keen memory that held every detail intact. To this day Jerry uses his dream world to uncover solutions to his problems or those of his clients. (He is also a computer consultant.) He dreams about these solutions holographically.

While most people discredit their nighttime counsel as "just a dream," Jerry has trained his mind to create his full presence in the experience. In his mind, he is there--seeing through his eyes, hearing through his ears and sensing and feeling with his body. What is amazing is that upon awakening Jerry can recall every detail of just how he is to build the device of which he dreamed.

Jerry knows that if he gets to a point where stress or frustration is building up, he can simply take a brief nap of ten to fifteen minutes, and invariably the solutions to his stress will come to him upon awakening.

I believe that everyone is imbued with some special talent or skill (genius). Sometimes the talent is obvious, such as the athlete or musician. Others are much more obscure--the born organizer, troubleshooter or mediator. Perhaps some just need more polish than others. Even people like Jerry, who clearly demonstrate a special gift, must choose to make it shine. Have you ever known an over-achieving athlete who went from being a "98-pound weakling" to captain of the football team? It was once said that geniuses aren't born, they are made. Although we can't all be like Jerry (nor would we want to be), we can learn from his methods to attain a sense of genius in our own right.

A genius understands the power within the *process* by which the five senses bring information into the brain. By using that power, the learning experience becomes a natural flow so that, just like turning a door knob, it requires little or no thought. This ability can hold true for any learned sequence. Even as a small child you had to start first by learning the muscles of the body and how to control them. Then you learned to crawl, then walk, and finally, to run. All of this was done through sequential or holographic learning, which uses all of the senses to link past experiences with future discoveries. This can be likened to a holographic picture that has been broken into a thousand pieces. With proper equipment the complete hologram could be reconstructed one thousand times over from each separate piece of the whole. All information is stored in every quadrant of the hologram.

Science has proven that the human brain's construction is similar to that of the hologram. The brain is one of the few parts of the body that is incapable of self-repair. Yet if a part of the brain is damaged, another part can be trained to take on the lost responsibilities or tasks. This relates to the theory behind genetic engineering which relies on the premise that *all* information must be stored in every cell of the body. Scientists believe that, once proven, it will be possible to take a cell from the human body, whether alive or dead, and reconstruct the physical image in a laboratory.

SEQUENTIAL LEARNING

Each megabyte of information currently stored in your mind was acquired through a process of **sequential learning.** Whether it is the skills of walking or running, asking for something or negotiating for it, all abilities are learned and processed sequentially.

People who have earned the title "genius" have trained their minds to work through the sequential learning process. For most geniuses of the past, it happened "by accident" (like my friend Jerry). Many of today's wealthiest and most famous inventors work in the field of computer electronics. What do you suppose would have happened if these powerful and creative minds were born into a different era? I believe they would have used these same skills of sequential learning, but for developing inventions appropriate for their time.

What about you? Have you ever considered how your sequence of learned experiences may have either helped or hurt your chances for achieving success? By understanding sequential learning, you will now begin unraveling the mystery of the genius mind.

There are two presuppositions to sequential learning.

1. **We learn in sequences or chunks of information.** Just like the example given above of how we learn to walk, the same is true in other subjects, such as mathematics. The student must first learn the sequence of numbers, then addition and the other areas of basic mathematics, before he or she can accelerate into calculus or any other advanced process. Even

in English, the student is first taught the Alphabet, then small words, phrases and finally complete sentences. If any one of these sequences is broken, the student may experience difficulties--perhaps throughout life, until a process is learned that will bridge the gap.

2. **The human bio-computer can learn at the rate of seven, plus or minus two, chunks of information at a time.**

Now it may seem as though this genius stuff is getting a little more complicated here, but it's really quite simple. Let's say you are a student. As you step through the doorway of math class, you hear the teacher speaking about algebra. Your first instinct is to pull a quick about-face. This is because your mind immediately starts to recall your belief that *you are no good at math.* That's one chunk of information. Now, depending upon your math ability, you are working with six, plus or minus two, chunks. Next, that negative thought triggers a memory of your father's voice exclaiming, *"You are never going to amount to anything!"* Now you are down to five plus or minus two chunks of information. This unhappy memory will most likely trigger another and then another until finally you would be left with no chunks. I'll bet you can guess what happens then. The self-fulfilling prophesy (your belief) would be correct--you *are* no good at math. You are now on a downward spiral as your mind successfully fulfills everything you believe in.

The mind can spiral in either a positive or negative direction--it's entirely up to you. In studies with my clients and in my personal experience, it has become apparent that when you are spiraling in a positive direction, you are operating at the top end of seven or more chunks of information--thus inspiring confidence and enthusiasm. If you are in a negative spiral, it seems there is an energy drain and you are operating at the lower end of the spectrum. If this is true, it would seem logical that your *present* state is quite an important factor. When you notice yourself spiraling into negative chunks of information, it is time to take action.

During the negative spiral you probably didn't use your whole brain, nor did you know how to set yourself up for a genius response. You will discover these options and more as you read through this book and learn how easy and natural it is to awaken your genius--your positive potential. You will open up the inherent genius part of your mind, that *other-than-conscious* mind that processes up to 20,000 different chunks of information per second. This part of you is like a child on Christmas morning, opening one present after another, thrilled with the *potential* in each unopened gift.

THE COMPUTER BRAIN

Your brain works like a computer in several ways. A computer operates on much more than just a control box, keyboard and a monitor. Yet these are the parts of which you are visibly aware. There are innumerable unseen internal parts which allow incredible mathematical functions to take place.

The key ingredient for any computer to function is its *operating system.* In computer jargon, this is usually a program called **D**ata **O**perating **S**ystem (DOS). Without this program your computer will not run. Your brain also has a DOS program and it consists of your *beliefs and values.*

Let's start with *belief systems.* One of my favorite statements is, *"The law of mind is the law of belief!"* What does this mean? Simply that whether you believe you are smart, or believe that everyone else is more intelligent than you, you are right! The mind will support you with the appropriate behaviors that will uphold your belief. This is why people who believe themselves to be unintelligent will sabotage their success in some way. If they do well on a test, they might say, "It was just luck." Then, on the very next test, they may fail miserably to bring their percentage back into a comfort zone. A genius is not afraid to stretch the comfort zone, to go beyond perceived limits, and is smart enough to know that *there is no failure, there is only feedback.*

Whether or not you consider yourself a genius, you can count on your brain never to make a liar out of you. If you enter into an activity with success in mind, your genius will continue to work out the details until you are successful.

You are what you think about all day long.

So what about values? The *value system* is known as a meta-program *(programs that work beyond the surface program and that rule the beliefs.)* Some people believe that values are learned through life's experiences and that once learned they are almost impossible to change. This is where the processes of the **Self-Help Dialogues** come into play. Values work beyond the conscious level and motivate us either to take action or retreat. Geniuses are capable of self-analysis. They will ask, *"Is there a better way for me to act, to think, to feel, to be?"* These are the people who take charge, are flexible and make changes whenever necessary. They learn early in life to cope with that which they have no choice over. A true genius manages and handles frustration.

Beliefs and values are very important as we move into the next area of discovery.

SELF-DISCOVERY: ARE YOU "WIRED" CORRECTLY?

Before moving on to the next paragraph, take a moment to stand up. Support your weight on your right leg and rotate the other leg in a circle. Continue moving your leg in a circular motion and spell your name in the air with your dominant hand.

What happened when you started to write your name? Most people find it easy to spin their leg in a circle, but as soon as they start writing their name the leg stops the circular motion and begins to move back and forth, or they are unable to write their name in the air. Whatever you may have experienced during the process, it provides strong evidence that there is an internal power struggle going on within you. This struggle is between the left brain functions and the right brain functions. A genius has learned to live in harmony with the right and left brain and has become whole in his thinking.

THE LEFT BRAIN

If you were a Vulcan, like Mr. Spock of Star Trek fame, you would probably have an incredibly active left brain. Like Spock, the left brain seeks logic; it believes in control and precision. The left brain is sequential and, therefore, everything in this area must make sense. Also, the *critical factor*, that part of the conscious mind trained to reject information before true consideration has occurred, is part of the left brain function. Whether you are verbally communicating with another person, working out mathematical equations, balancing your checkbook or solving a word game, you are using the left brain. Today's school systems operate primarily with left brain processing through memorization. That is why some students find school to be a big YAWN. These kids are usually the "Right Brainers." *(We'll discuss them next.)*

Unfortunately, there is one small glitch with the over-active left brain's logical and linear way of thinking--it assumes that all information is 100% accurate just as it stands. No problem, unless you were trained with improper or inappropriate beliefs and values. If this is the case, the logical thinking of the left brain can end up controlling you with negative self-fulfilling prophecies (like the math student who was *"no good at algebra"*). It is certainly no coincidence that most addictions or problem behaviors stem from left brain functions. The person with an over-active left brain locks out the creativity that could otherwise present an optional behavior or alternate method for success.

The Left Brain is not concerned with what is true; it is only affected by what is truth *for you.* Your truth is based on the evidence you have gathered throughout your life experience, whether real or imagined. This information is your judge and jury. It controls and plans your future because time is also monitored by the left brain. A genius realizes that true power is in neither the left nor the right brain functions but in the ability to fluctuate between them according to what is needed at the time.

Now don't get me wrong, the left brain functions are a necessary part of day-to-day thinking. My hope is that by now you are enlightened to the possibility that maybe, just maybe, the way your left brain has been trained has caused it to stand in the way of your success.

When the left brain is equipped to work in conjunction with the right brain, harmony will occur. It is in this state of harmony that optimism and hope can flourish. Einstein once said that the brain works like a parachute: it works best when open! If the left brain is closed to new ideas, the powerful flow of the mind will cease.

I will be guiding you through a learning process of discovering how to create a truly genius mind, one that will allow you to flow in and out of your left brain and use your *other-than-conscious* mind to create *solutions* in your life. The average Joe sees life as a series of problems, while a genius sees life as a series of discoveries and solutions.

The True Power Of Your Mind Does Not Reside In Knowledge.

I am a great admirer of Albert Einstein. It is obvious that Einstein knew how to use his brain to get results. I even have a giant-sized poster of young Einstein hanging in my office. I bought it not so much for his picture, but because of a simple yet profound quote which is printed along the side of his face. He tells us, *"Imagination is more important than knowledge."* I believe it was this simple wisdom that led Einstein to his greatest discoveries. If you really want to open up to greater possibilities, you must open up your creative right-brain functions to work in harmony with your logical mind.

As I researched the many geniuses of our time and of long ago, I was amazed to find that a good number of the great inventors of the past were also artists. Leonardo da Vinci, one of the greatest artists of all time, was also an extraordinary inventor. Leonardo had a curious mind, and he often directed his artistic talent toward scientific endeavors. He used his drawing skills to convey his scientific conceptions. Most of his inventions, however, were much too advanced for his generation. His drawings revealed what he conceived possible for the future: flying machines, parachutes, helicopters, underwater diving suits, a protective tank for battle and methods of automation, to name just a few. The tools and known resources of the period could not come close to producing that which his imagination could so clearly create. Most of Leonardo's inventions were unknown until the 20th Century, when several of his notebooks were found. Interestingly, these notes were all written in a bizarre right-to-left script which could only be read with a mirror. We can only speculate as to his purpose in writing the notes in such a peculiar manner. Yet it is clear that these are the writings of an incredibly creative mind.

The average, everyday students of today are incredible geniuses in comparison to past generations, simply by virtue of the volumes of information available to them. It is estimated that you encounter more information in one day than your great grandparents did in a lifetime. Through the media, books, radio, television, computers, billboards, and shopping malls you are bombarded with an incredible amount of information each day. Scientists and psychologists speculate that if a 19th Century cowboy were placed in modern-day Los Angeles, his physical, mental and emotional state would be so severely affected that he would probably not survive more than a month.

It is mind boggling to think of the amount of information our children absorb, store, categorize and use each day. Clearly, by any standards, these kids have remarkable genius minds. Unfortunately, these incredible minds are being programmed by television sitcoms, commercials and video games. If you want to be convinced that children have perfect memory and recall, just listen to them during idle time when they resort to repeating all the commercials they have seen and heard. How do they remember advertising so vividly? It is because the ads are bright and colorful with movement and action. They are set to music, catchy and continuously repeated. The people who have products to sell are programming our children.

What is it like when a child goes to school? The information is usually dull, presented in black and white text, with the monotone voice of the same teacher day in and day out, and the work is often redundant. Many of the teachers have themselves grown bored with the material after presenting it year after year and do little to add sparkle to their presentations, let alone update their information.

The personal computer first made its appearance in the 1970's. Today many households have PC's and almost every school system has some form of computer available. Within a thirty-year time span all major corporations and most small businesses have become computerized to some degree. But has the computer age been integrated into the educational system? Is the computer a classroom tool, or an oddity that sits in a separate room down the hall, where each child may get an hour of instruction each week? Ask today's kids which class is their favorite. If they are fortunate enough to attend a school with a computer program, it is usually at the top of their list. These kids, ready for the 21st Century, have the dynamic minds of the computer age, but they are stuck in a dull, windowless classroom adding two plus two.

Are you ready for learning to come of age? Then read on!

The Right Brain

When I was a child I used to marvel at the mind of the inventor. How did these geniuses ever come up with their original ideas? Why hadn't anyone ever thought of it before? How did they know it would really work? Was their brain any different than mine? Their heads didn't look any bigger than mine or anyone else's. I once read that Albert Einstein didn't even graduate from high school, and he actually failed an examination that would have allowed him to pursue a career as an electrical engineer. So how did he become the man whose name is synonymous with genius?

As I grew up and made my own discoveries, these questions never completely left my mind, and, to be honest, the answers were a long time in coming. I was curious, and it was this sense of wonder that attracted me to the fields of psychology and counseling. I have since drawn many of my own conclusions and this is one of which I am certain: Those people possessing right-brain dominant features have a better opportunity for making changes and becoming balanced in their thinking.

The right brain is responsible for our ability to be creative; it is the ability to dream. The musical and artistic skills are products of the right brain. Some consider these features to be more feminine in nature, although men and women can be equally creative and artistic. The right brain is a wonderful, free-spirited place to live. It is the place of dreams and visions, fantasies and fairy tales, freedom, imagination, romance and make-believe. It is a place that knows no boundaries or limitations.

Sound wonderful? It is. But for some, it can be a dangerous place too. When people allow themselves to become too right-brain dominant in this left-brain world, they are setting themselves up for trouble. How long does one last in our present society with the inability to balance a check book, or to perform simple mathematics, or stay focused on the simple laws of English?

It is unfortunate that our society has evolved into a framework of left brain dominance. What is truly sad is that our school systems have followed suit. I suppose this has happened out of necessity. No one wants to raise a child to be a "dreamer" when society demands that everyone make a living, have a nice home with two cars in the driveway, keep money in bank accounts, and carry an array of little plastic cards in their wallets. This *is* the American Dream, *isn't it?*

If our school systems don't move beyond their nineteenth century foundation, they will continue to churn out honor-roll students based on who was able to develop the best memorization skills. Those who excel in music or the arts will usually muddle along in the classroom until their left brain functions mature. In many countries, such as Russia, these special students of the arts are placed in programs specifically designed to develop creative skills and to perfect their natural abilities. What would the world have missed if one special little boy was not allowed to develop into a great composer? Mozart began writing minuets at the age of five and, although he lived only 35 years, he composed 626 catalogued works. Mozart is considered a genius in music, as is Leonardo da Vinci in art and Albert Einstein in physics.

So how *did* Albert Einstein become known as the greatest scientist of our time? Well, in 1896 he was finally accepted into the Zurich Polytechnic and later graduated as a secondary school teacher of mathematics and physics. His first two years after graduation were lean ones, but he finally obtained a position at the Swiss patent office. The work was tedious and demanding, yet while employed there he completed an extraordinary range of publications in theoretical physics. These texts were all written in his spare time and with no outside contact with the scientific community or literature of the time. He later submitted one of these scientific papers to the University of Zurich and was granted a doctorate degree on its merit. The next year he was granted an appointment as associate professor of physics at the University of Zurich. In the short span of thirteen years Einstein progressed from a primary-level education to recognition as a leading scientific thinker. What was his secret? He knew what he wanted and was determined to get it!

So what excites you? Music perhaps, or art? What about the sciences--physics, biology, chemistry? Perhaps it is the written word--English, writing or poetry? Is there an actor within you just waiting for a moment to shine? Do you have a comedian hiding inside, ready to leap upon a stage? Perhaps a dancer, athlete or mime? Is math or history your thing, perchance? Are you fascinated by human beings? Perhaps you are a people-watcher--what about psychology, sociology, or anthropology? Whether it is one of these fields that entices you or another is of little consequence. What is important is that you find what you enjoy and *do it*.

THE RULE OF AWAKENING YOUR GENIUS . . .

is to attain the skills of self-discovery. You must first become a genius *about you*! Self-discovery is to know yourself and to truly own your successes *and* your faults.

> ***"Know Thyself."***
> Inscription -Temple of Apollo

Naturally, you will need a starting point. Every one of us possesses many wonderful talents, skills and abilities. These are as varied as people themselves. So how do we start the process of tapping into them? First, we must understand who we are. Think about it; right this very moment there is an internal genius at work within you. It is somehow monitoring the seventy-five trillion cells of your body and replacing those cells at the rate of fifty million or more per second. Your eyes are scanning across this page and, believe it or not, your super-conscious mind has already absorbed every word that is printed here. It is all happening automatically and with incomprehensible precision and accuracy. Sheer genius!

So why don't you feel like a genius? Because the limits of your conscious mind have never been expanded. Everyone has the genius potential. It is common to all of humankind. Something greater than you controls your physical, mental and emotional nature. All you need to begin the process is to get out of your own way, no matter what the shame, guilt, resentment or other negative experiences of the past may tell you. It is time to unlearn any past programming from your parents or family members that is not working for you. Keep in mind that your parents were trained by their parents, who were trained by their parents--and not one of them was ever given a "Parent Training Manual." They did the very best they knew how with the information made available to them.

I consider myself fortunate to have grown up the son of an alcoholic. I had the benefit of many revelations very early in life. I was always trying to understand my father and his behavior. Why couldn't he just say "no" to the booze? Then one day I realized that perhaps I would not respond much differently than my father if I was under the same kind of stress without any training or guidance. This awareness was a key step for me in awakening my own genius.

I could not control another person, only accept him as he was and take steps in the direction of my own goals. As I accomplished goal after goal with this in mind, the anger and resentment that I had previously held toward my father somehow vanished. It has been my experience when counseling others that the same can also be true of any relationship; with parents, siblings, teachers and even close friends. The truth is that we are *all* far greater than we have been led to believe and far greater than most people can consciously comprehend. If you are experiencing a relationship that just isn't working, imagine what might happen if you simply released the need to control the other person, or the situation, and began focusing on more positive endeavors. Try it! I guarantee that you will get a new and better outcome.

THE OTHER-THAN-CONSCIOUS MIND

There is an available power that is greater than either the right or left brain functions--and we can identify it as the *other-than-conscious*-mind *(the genius mind).* This power needs to be used in a practical way and, when used effectively, works as a flow or channel from the right to the left brain and back again.

Everyone has some degree of creativity. In truth, as humans we are in a continuous state of creativity--either consciously or unconsciously we all create the situations that are present in our lives today. If we are honest with ourselves, we know what the end result of our actions will be, whether positive or negative. If we take the act of living to its simplest components, we can see that life is a journey; an unending series of problems and solutions. The genius realizes this and doesn't get stuck anywhere along the way, but rather flows with the experiences of life. There is a part of us known as the "*creative part,*" which is the part that can discover and implement a solution to any given problem.

Understanding this *creative part* of you will be a big step because, after all that I have just explained, I am now going to tell you that you don't really have a right brain and a left brain. Research has proven that either brain is able to perform all of the functions of both. We have simply trained ourselves to use one or the other. If a person damages either the right or left brain, the unharmed brain can be trained to re-learn the lost functions.

So, guess what this means? You have two brains . . .

You have two bio-computers or brains which are joined by the corpus callosum, a strip of gray matter comprised of a band of nerve fibers. A genius knows how to use his or her mind to create what is known as ***whole brain thinking***. You can think of whole brain thinking as the ability to use both hemispheres equally; *controlled* **creativity** or *precise* **dreaming**. When accessing the whole brain, you are able to master verbal and nonverbal communication, develop artistic abilities, and take in all information with perfect memory and recall.

The processes in this book are designed to awaken you to the full potential that resides within you. Avoid falling prey to the belief that genius is what some lucky somebody was born into. We all possess the ability to be a genius in our own lifetime; so ask yourself:

- What would life be like if I could turn on creativity whenever I needed it? ...*Controlled* **Creativity**.

- What would life be like if my dreams were clear and precise and I would know just what I need to do when I need to do it? What if I could begin to live my dreams? ...*Precise* **Dreaming**.

- Did you know that the music of Bach and Beethoven balances the brain and allows it to go into the Alpha state, the best state for relaxation, peace of mind and creative solutions and ideas? ...*Logical* **Music**.

The list could go on and on. Can you imagine doubling your ability to remember and recall information? It is with this possibility in mind that I want you to move forward in reading this book. To help you put the sequence together so that whatever you want to accomplish... whatever your dreams... you will be in harmony with them. And, once and for all, allow yourself to *be* the genius that is you.

SELF-DISCOVERY: STRETCHING THE RIGHT/LEFT CONNECTION.

An easy way to exercise whole brain thinking is to stretch its capacity. Hold a crayon or marker in each hand the way you would normally hold your pen. Place the points of each together in the center of a piece of paper. Starting from that center point begin writing the word "elephant" moving in an outward direction with each hand. The left hand moves to the left, and the right hand flows to the right.

If you found it difficult to do this process, keep trying. Remember practice makes perfect. Here's a hint: Think of your non-dominant hand as creating the mirror image of your dominant hand.

Left hand - Right hand

elephant elephant

THE BRAIN IS LIKE A MUSCLE

The brain is like a muscle. The more you use it the stronger it gets. You can think of the exercises in this text as *mental fitness*. Remember that the brain needs to be whipped into shape, so at first the exercises will seem simple. But we will soon move into more complex mental exercises.

SELF-DISCOVERY: FINDING YOUR DOMINANT BRAIN.

Is one or the other side of your brain dominant? Probably. Since you were never trained to keep both sides active, one will almost always take dominance. Here's a great way for you to find out.

Face the corner of a room. Hold your arm straight out in front of you with the thumb pointing up. Place your thumb directly in the corner of the room. Then close each eye, one at a time. Find out which eye, when closed, allows the thumb to stay centered in the corner.

Are you right or left brain dominant? In most cases, if your thumb stayed centered with your right eye open, then your left brain was dominant and if your left eye was open then your right brain was dominant.

The true purpose of this exercise is to help you understand the way in which *your* mind processes information. Would you believe that when your eyes take in the world around you they immediately flip the image upside down and then show it on the screen of space behind the other eye? Remember the Self-Discovery, "**Are You Wired Correctly?**" where you made a circle with your leg and wrote your name in the air? This is a perfect example of how the brain is cross-wired. If you are using the right side of your body *(eye, ear, arm, leg),* it is your Left Brain that is processing the information; and if you are using the left eye, ear, arm or leg, it is the Right Brain doing the work.

It is with this understanding of the brain that you can now move on to the next stage of learning about this incredible person that you are.

CHAPTER TWO

LEARNING WITH STYLE

The functions of the right and left brain are only a small portion of the potential that comprises *the mind.* The mind and the brain are two very different things. You can think of the brain as the hard drive of the computer and the mind as the software. Your mind is the intelligence that is in every cell, every system and every organ of your body. It is that intangible, incredible, and awe-inspiring part of you that thinks, remembers, sorts information, makes judgments, forms habits, discerns, distinguishes, communicates and otherwise makes you who you are.

Your mind takes in all information through the five senses (sight, touch, hearing, smell and taste). You can liken the five senses to the keyboard on the computer. Through the senses you receive input; it is your data entry system. Your brain then stores, categorizes and utilizes the information. A genius knows how to access that information and then apply it in discovering the solutions to any problems that may arise. It is therefore essential that you discover your own individual style for gathering and communicating information.

SELF DISCOVERY: FINDING YOUR PREFERENCE

As you read through each of the following questions, think about your own likes, dislikes, habits and behaviors. Then use the scale to choose the appropriate response to each statement and write the number in the space provided.

I Agree = 10 **I Disagree = 5**

1. ______ My friends know me as a good listener.
2. ______ I prefer movies with a relaxed pace and a plot that slowly unwinds over those that are full of action, noise and special effects.
3. ______ My idea of a great evening is when I can just stay home and wear comfortable clothing.
4. ______ I am inclined to make my first impression of people by the way they use their voice rather than by their looks or physical actions.
5. ______ I love to watch the people go by at the shopping mall.
6. ______ I have a vivid imagination.
7. ______ I can hardly resist singing along with the radio whenever it's playing.
8. ______ I won't leave the house until I am certain that I look good.
9. ______ I can truly enjoy music only when it helps me to relax.
10. ______ I will go to certain movies just to see the "special effects," scenery or costumes.
11. ______ There is nothing that can relax me more than having my neck and shoulders rubbed.
12. ______ I spend a good deal of my leisure time on the telephone.
13. ______ I need to get up, stretch and move around frequently.
14. ______ After a stressful day my body will feel tense and I frequently have a difficult time unwinding.
15. ______ I usually watch television or read while eating.
16. ______ I would rather listen to a story teller than read a book.
17. ______ I have a clearly defined concept of what I want my life to be.
18. ______ I regularly listen to talk radio.
19. ______ I will usually judge people by their clothing and appearance more than by the way they speak or physically respond.
20. ______ I usually fidget or doodle while talking on the telephone.

21. 5 I will often spend time just listening to my tapes or CD's.
22. 10 I find it difficult to tune out noise and loud voices.
23. 10 I find it easy to hug someone whom I have just met.
24. 10 I will usually wait for my "gut feeling" to decide how I feel about a person.
25. 10 I am drawn to books with attractive, colorful covers.
26. 5 To me, there is nothing so stimulating as good conversation.
27. 10 I notice the decor and artwork when entering a room.
28. 5 I sing, hum or talk to myself while taking a shower.
29. 10 My bedroom is color-coordinated and well decorated.
30. 5 For me there is nothing like a hot bath to relieve stress and tension.

Now use your response totals from above to figure out which mode of communication is your preference:

VISUAL		AUDITORY		KINESTHETIC	
Question Number	*Response* Total	*Question* Number	*Response* Total	*Question* Number	*Response* Total
5	10	1	10	2	10
6	5	4	5	3	5
8	5	7	5	9	5
10	10	12	5	11	5
15	10	16	5	13	5
17	5	18	5	14	10
19	10	21	5	20	5
25	10	22	10	23	10
27	10	26	5	24	10
29	10	28	5	30	5
TOTAL	85	TOTAL	60	TOTAL	70

HOW DID YOU RATE?

When adding up the scores, did you find a balance between the totals? If not, don't feel alone. Most of us use one sense more dominantly than the others. In the same way as we tend to rely on one side of the brain, either left or right, we also tend to allow one sensory system to dominate how we perceive our world.

Go back over the list and review the questions to which you marked lower scores. Think through the questions once again. Would you be willing to start stretching your abilities by consciously choosing to use these other senses more? Can you imagine opening your mind to receiving *all* the input from the world around you?

It is natural for you to use what is comfortable, but the awakened genius is willing to grow. Like the turtle, you won't go far without coming out of your shell. You will be pleasantly surprised as you practice using your other senses. You will notice how your memory improves and how easy it becomes for you to make choices. You will discover a whole new world of communication. Once you learn to create a balance in perception, you will be well on your way to thinking holographically. In the following section you will gain a greater understanding of each sensory filter.

VISUAL, AUDITORY AND KINESTHETIC PROCESSING

Now let's find out what all of this means to you, the genius. As you read through the remainder of this chapter, think about yourself and also your friends, family members and teachers. Notice the way these people might prefer to learn and communicate.

We all have preferences for how we like information to be presented.

*Some like to **SEE** what you mean .... **VISUAL***
*Some like to **HEAR** your idea ...... **AUDITORY***
*Some like to **EXPERIENCE** or **FEEL***
*what you are talking about .... **KINESTHETIC***

Similarly, we also have preferences for the way we evaluate and analyze information:

Some decide by how things **LOOK** *to them . . . .* **VISUAL**
Some decide by how things **SOUND** *to them . . .* **AUDITORY**
Some decide by how things **FEEL** *to them . . .* **KINESTHETIC**

Generally, people will take in information and communicate with all three of these access modes. Some will have one kind of preference for gathering information, and another kind for interpreting it. For example:

LOOKING at a new car and then buying it because it **FEELS** right.

You will soon be given the opportunity to better understand how your mind processes information. As you read through this section, keep in mind that developing the flexibility to use all three is what will awaken your genius potential.

THE VISUAL LEARNER

When you go to your closet to get dressed each morning, what do you find? Is your clothing bright and colorful and always coordinated? What about your grooming practices? Do you spend a great deal of time in front of the mirror making certain that your appearance is just right? What do you notice first about other people? Is it their appearance, clothing and grooming? Are you always consciously aware of your surroundings? Does a tilted piece of art work or a misplaced item of furniture drive you to the edge?

If you found yourself nodding your head in acknowledgment, then you are probably a visual access person. Studies have shown that 70% of the U.S. population is made up of visual processors. This can be directly attributed to the American culture which is filled with visual entertainment, bright lights and flashy clothing. Most visual learners must *see* what they are to learn or they simply do not absorb the information. Without a visual stimulus, retention will drop to half or less. Because of the high percentage of visual learners, our school systems operate on a visual premise. There is a presenter (the teacher), using visual aids and reading material to convey the lesson at hand. The addition of televisions and VCR's in the classroom have expanded the visual process.

THE AUDITORY LEARNER

The auditory learner may not be quite so concerned about appearance, but is very concerned about what other people have to say. *Will he tell me that I look pretty? Will she tell me that this shirt shows off my biceps?* The auditory person may not care too much if his car needs a paint job, or if the interior seats are slightly worn or frayed--but, WOW, what a sound system! An auditory buyer would look for an automobile that is ***soundly*** built. He may open and close a car door several times just to hear if it sounds right.

Auditory learners are the people who can sit through an entire lecture without ever looking at the speaker, but silently absorbing all that is being heard. Sometimes people will assume that an auditory person is not listening because he or she will rarely look directly at the speaker. In reality, auditory learners are completely tuned in to what they are hearing and are using very little of their visual sense.

THE KINESTHETIC LEARNER

Remember those kids in your class whose clothing seemed to hang from their body and never fit quite right? You know the kid, the one who always wore those frumpy sandals, the comb never seemed to make it all the way through her hair, and her glasses rested at a tilt more often than not. Certainly not every kinesthetic learner is *this* "comfortable," but comfort tends to be their top priority. The kinesthetic learner accesses the world through feelings. These are the people who will reach out and touch your hand or shoulder while talking with you. They love to hug and feel close and warm with other people.

To the rest of the world, the kinesthetic learners will often seem slow, or even dull. This is because they need *to "get a handle"* on the information and then *"get a feel for it,"* before they can make a decision. Most kinesthetic learners need to use their neurology (body) to process information. They are often fidgeters and work best when they are able to use their hands. All of this emotional processing takes time, and often the visual and auditory learners are off on another tangent long before the kinesthetic learner has "grasped" the concept.

Our school systems are not designed for either the auditory or kinesthetic learners, but these "feelers" tend to be left with the greatest handicap. The auditory person at least has the benefit of the lecture that goes along with the visual presentation, but the kinesthetic processor is told that she must be *still*, *look* at the board and *listen* to the presenter. These guidelines all lock this learner out of her natural genius. If only they would give her something to *do*.

THE BALANCED GENIUS

Everyone learns and communicates in all three of these modes. However, most people develop a preference and tend to use one access much more frequently than the others. Now go back and review the results of your Self-Discovery Exercise, **Finding Your Preference**, and begin to think about how your personal communication preferences have been affecting you.

As an awakening genius it will be your goal to become aware of your own preference and then strive to build a balance between the three. All of the Self-Help Dialogues are specially designed to create this sense of harmony between the visual, auditory and kinesthetic processing that occurs within you. Although the dialogues are an auditory process, with your eyes closed and with a comfortably relaxed mind you can imagine seeing, hearing and feeling yourself successful. Then, when you are out in the real world creating success, you will know when you have arrived; you have already visualized, listened to, and experienced total success!

Remember that your mind is holographic, it is always accessing with all five senses. I have a brother, Walter, who has a unique way of saying good-bye. He will say, *"Smell ya later!"* This seems odd to most people, but to him it seems perfectly natural. This is because Walter tends to be more tuned in to his olfactory (smell) sense than most people--so he uses a language that relates. The same is true with most people--they will generally communicate with their preferred mode.

Conversations And Miscommunications

Just as you are now developing an understanding of yourself, you will also begin to relate better with others. As you read over the following dialogue between a teacher and student, pay close attention to the modes of communication being used. Are these two communicating successfully? Do you identify with one or the other?

TOM TEACHER: *I want you to LOOK over this test. Can't you SEE the mistakes you are making? How do you expect to get anywhere in this world when you don't have any VISION?*

SALLY STUDENT: *Well, I really FEEL that I did my best. Somehow, I just can't seem to get a HANDLE on how to study for tests. I'm just not GRASPING the concepts.*

TOM TEACHER: *I don't think you're taking the time to SEE the entire PICTURE. I realize it all may APPEAR somewhat VAGUE to you right now, perhaps if you will let me SHOW you the process one more time.*

SALLY STUDENT: *I don't know if that will help. I need to GET IN TOUCH with my FEELINGS on this. I just don't FEEL COMFORTABLE in this class. I'll TOSS IT AROUND in my mind and get back with you.*

Will TOM convince SALLY to allow him to "SHOW" her the work again? Who could benefit from improved communication?

It is not likely that TOM TEACHER will convince SALLY to "SEE" his "VISION." Indeed, chances are good that SALLY will pull away from TOM TEACHER and the class even more.

Did you get frustrated with TOM? How about with SALLY? Which of these two could benefit most from improved communication? Both. It is TOM's job to teach SALLY the information required for his class. If TOM had spoken in SALLY's language (kinesthetic), he may very well have been able to help her to "GRASP" the concepts.

SALLY, on the other hand, has a duty to herself to learn what is presented in the classroom. If she had been willing to communicate in the language of her teacher (visual), she would have allowed herself access to his "VISION" -- the information needed to succeed.

The genius is keenly aware of the power that is communication. Understanding an instructor's communication preference will put you a step ahead in the classroom. Relating to other people with their preferred access will give you the leading edge in any conversation.

Take this knowledge out into the world with you. Begin paying attention to the communication preferences of your neighbors, friends and even strangers. Maybe that next-door neighbor who never so much as glanced in your direction when you greeted him with a cheery "Hello," is, in reality, an auditory wizard, absorbing every detail of the sound of your voice! And what about Ms. Perfect, with her impeccable appearance and neat, attractive life? Perhaps she is simply accessing her visual genius potential! Will you have a new opinion of that grocery clerk who always seemed so slow? Perhaps he is really a kinesthetic genius!

"Intelligence is not the ability to store information, but to know where to find it."

Albert Einstein

*"Let us learn to dream,
gentlemen, and then
perhaps we shall learn
the truth."*

August Kekule, 1865

CHAPTER THREE

YOUR RELATIONSHIP TO THE PAST

"If you look in the rear view mirror for too long,
you will crash."
Jim Ward,
High School Football Coach

People tend to spend a great deal of time dwelling on the negative experiences of the past. This serves no logical purpose. The brain is a goal-oriented organism. It only knows how to succeed. Therefore if you are concentrating your attention on negative experiences, your mind will strive to succeed at producing more of the same. Have you ever known a person who seems to have absolutely the worst luck in the world? How did this person spend his or her time? Was it in talking endlessly of sadness and woe? Was the next calamity just around the corner? Could it be possible that this person was creating his or her own bad luck?

My eighth-grade school year was an educational nightmare. The classroom had been an incredibly unhappy place for me ever since my second-grade year, when I was held back and labeled "learning disabled." About the same time that I was skimming the surface of eighth grade, my father was actively overcoming alcohol addiction. He had replaced his thirst for booze with a craving for knowledge and his search led him to the concepts of "positive thinking." He was eager to pass his new-found wisdom on to his children.

I found it very difficult to understand how this "mind stuff," could really work. At that tender age, I had already developed a very negative attitude about people in positions of authority. I can remember thinking that my teachers' only purpose in life was to cause me grief and that my floor hockey coach *must* know less than nothing. I had the habit of starting a new school year with the firm conviction that the teachers would look at me just to find fault. Invariably I would set myself to the task of proving them right. "Trouble" became my second name. I spent more time in detention than in the classroom -- I had to live up to my reputation! By the time I was twelve years old, I found it impossible to imagine that my future could be any different or better than what I had "always" experienced in the past.

One day my father took me aside and asked me to sit down beside him. I prepared myself for either an endless lecture, being grounded for life, or one of those father/son talks that go nowhere; or perhaps all of the above. But my Dad, because of the many difficulties he had overcome in his life, was now able to help me unravel the mystery of the mind and to free myself from such negative thinking. He pointed out that there is one common denominator in each of my unhappy school experiences. The teacher, classroom, and material would change from year to year -- what remained constant was ***me***!

It all began to make sense; I had been allowing the future to be created out of my past. It was not my teachers or my environment that needed to change, I was the only one who could break that past model. I simply needed to think, act and, most importantly, *respond* differently.

According to my past, I was a dummy. I had been branded "learning disabled" and was held back in second grade because of incredibly poor reading and language skills, a pattern that progressed throughout grade school. It took me a whole summer of contemplation and several years of added effort to break free from that past mold. During my sophomore year of high school I reached honor-roll status, a position I maintained all the way through college. If I had continued the past pattern of creating my future by following the route of my past, it would be impossible for me to be writing this book today.

This next section will show you how, by taking control of your thoughts, you can make changes happen in your life instantly and automatically. Take your time and think through each of the questions presented to you. Think of the way you would get to know someone who interests you, and truly get to know yourself.

> *"There is the greatest practical benefit in making a few failures early in life."*
> **T.H. HUXLEY**

WHERE ARE YOU?

What is your relationship to the past? Think of the types of behavior that you display on a daily basis. When do you do them? What triggers them into action? Are they truly yours? Or do you think of your mother or father, or perhaps a friend or other relative in relation to these actions? If yes, are these behaviors positive? If so, keep them! If these are negative or unwanted habits, however, you can use the patterns that follow to either rid yourself of them, or transform them to what you would like them to be. You need to know where you are so you can find out where you are headed.

At times life can be something like being aboard a large jet airliner. We assume that the pilot (the conscious mind) knows where he is going and how to get there. But, if the going gets rough, we may start to feel that we have given a little too much of our power over to the pilot. Maybe we didn't even give him the correct flight plan! There are many reasons for finding ourselves off course, and they may be real and valid, but there are more and better reasons for getting back on course.

I recently attended a conference in Atlanta, Georgia. While there I had the opportunity to hear an intriguing story told by Thomas Grinder, an inspiring motivational speaker. The deeper meaning and implications of his story stayed with me for days after I heard it, and I still get chills each time I repeat it. I believe it to be one of the greatest metaphors I have ever heard relating to the flow of the mind. Since it is actually a metaphoric story, I don't think Mr. Grinder will mind if I repeat it here.

Mr. Grinder's fictitious friend, we'll call him "Caleb," has a theory that when he is in the flow and presence of life (in harmony with the *other-than-conscious* mind) it's as if life itself is carrying him all the way to his goals of the day. These are the days when he feels confident and capable. He knows his direction and the path is clear.

Caleb noticed, however, that on those days when he is, for some reason or another, not in the flow, it feels as if he is trudging through muck and mire clear up to his thighs. On one particular day when that muck felt so thick and his legs so heavy that he was sure he could go no further, he finally looked wonderingly up to the sky and asked a simple question. *"Where are you, God?!"*

Caleb gazed steadfastly up to the heavens, but no reply was forthcoming. Finally, he decided to press on and, with a heavy heart, began to wade again through the mass that now felt like lead weights drawing him deeper and deeper. Then, out of the corner of his eye, he noticed that someone was waving to him. He looked closer and then recognized the form. *"Dear God!"* he cried. *"What are you doing way over there when I need you here?!"*

"I'm over here," God called back, *"because this is where the path is!"*

Now, I don't know whether God really manifests himself in this way, but I do know that God works in mysterious ways. There is no possible way for us to comprehend the plan God has for our lives. Sometimes His plan can seem like total distraction, but after the storm and confusion settles, everything works out. Have you ever thought about a past experience and then wondered to yourself, "What ever made me think *that* was such a big deal?" Didn't it all work out for the best? Awakened geniuses know there is something greater than their little human self keeping this world together.

As an up-and-coming genius you will want to walk the path of least resistance. If life feels like muck and mire, it's time to seek a new path that is clear, straight and will lead directly to your goals.

Are you perhaps a little like Caleb -- intent on treading a path of muck and mire? If so, there is no time like the present to look around you at all the options that are available. Options are endless to those willing to change. It might seem like a bizarre thought to some of you reading this, but *life was meant to be FUN.* Many people get confused by thinking this means that *life is supposed to be EASY.* Sorry folks, this is a do-it-yourself world. Once you wake up your inner genius to this simple wisdom, it will surpass all other knowledge; it will lead you down life's most direct path as well as help you to get back on track whenever you wander off. Although money is often used as a yardstick for success, it doesn't always guarantee an *easy* life. What most people are truly seeking is a *feeling*--a happiness or peace of mind--that will only come when they believe it to be so.

It was the last month of my sixth grade year--one of those mild spring days when every child in the school is particularly restless and boisterous; anxious for the freedom summer promises. The teacher tried her best to settle the class, but to no avail. To one overly excited boy she finally exclaimed, "Get busy on your work or you'll end up nothing more than a garbage man!"

Her hands were planted firmly on her hips and her toe was tapping impatiently. "Is that what you want?" she prodded.

Her comments were not all that successful; it was a threat we all had heard many times before.

The next day another boy from my class arrived with a tall, burly man who had shoe-leather skin. It was his father. His huge work boots thunked out his steps as he strode across the room toward the teacher, "Listen," he said, pointing his finger at her face, "It just so happens that I make my livin' pickin' up garbage for this here city and I really don't care for the way you been knockin' my profession to these kids. Why, I make more money than you do!" He was blustering now. "I'm a good Dad... and I support my family. We got a nice house and car... and it just so happens I like my work!"

It would seem this fellow had a much different perception of "garbage man," than my teacher did. In his "do-it-yourself" world, collecting garbage was an honorable profession and he considered himself rather successful. Unfortunately, there are probably many other garbage collectors who do *perceive* their job as degrading. Because this is their perception, they are right!

Now that you are learning to recognize the power within your own genius (*your perceptions*), you won't need to get so far off course to make corrections. Alcoholics Anonymous has a special statement that helps their members stay on course. I find it valid for every human being. *"God grant me the courage to change the things I can, the ability to accept the things I cannot change and the wisdom to know the difference."*

Sammy Sly has a goal and is certain that one day he will achieve it. He knows he has set his sights high and the neighbors often ridicule him for trying, but still he is determined--one day he *will* see the Statue of Liberty. So every morning the buzz of his alarm awakens him at sunrise. He tidies his little two-room apartment, then showers, shaves, and eats a healthy breakfast. With upturned chin and light feet, he steps out into the bright sunshine, savoring the warmth and ignoring the smog of the day. He sets a straight course for the bus-stop. Forty-seven minutes later he steps down onto sand-covered pavement and draws a deep breath of the salty ocean spray. Sammy spends his day combing the shoreline, ever hopeful that today he will discover the majestic statue of his dreams.

At mid-afternoon, Sammy's confidence becomes overshadowed by a familiar sense of gloom. He then stops, stretches and gazes down the expanse of open beach. Off in the distance he can see the Santa Monica Pier. With an ache in his chest he turns to look upon the expanse of city behind him. He can just glimpse the highest peaks of the Hollywood Hills. At sunset Sammy boards the bus for home. He climbs the steps with leaden feet, slumps into the seat with drooping shoulders, and lets the hot tears of frustration stream down his flushed cheeks.

Will Sammy ever find the Statue of Liberty? Of course not. He is determined, steadfast, and absolutely certain that his goal will be achieved. Yet he is 3,000 miles off course. A genius doesn't jump out of bed every morning and run to the west to see the sunrise. He wakes up with the knowledge that everything he needs will be obtained through this journey called life. If he is lacking the ability he doesn't cry, *"Why Me?"* He takes inventory of himself and then asks, *Why not me?* He can then set out on a new course that will provide any additional help or training he might need. Consider what Thomas Edison once said, *"Genius is 1% inspiration and 99% perspiration!"*

SELF-DISCOVERY: DRAWING YOUR RELATIONSHIP TO YOUR PAST

It's time to set your course. In order to determine the best direction to take, you must first know where you've been. That's right, get out paper and markers or crayons, let your right brain take over, and draw your relationship to the past. You can use colors or symbols or anything else that may reflect how you now relate to your past.

Did you take the time to draw the picture of your relationship to the past? If you passed that exercise by, go back now and draw your picture. It's important because, believe it or not, most people still think their past controls them. Nonsense, you might say, my past can't control me! Well maybe not consciously--but on an unconscious level your past may have a firm grip on you.

As an example, once when "Nancy" was very young she did terribly on a spelling test. She felt embarrassed and ashamed and grew extremely fearful of the next time she would be quizzed in spelling. She then began to build the belief that she was a horrible speller. With each testing situation her fear became stronger, and her belief in herself as a poor speller was affirmed. Now Nancy may have had the ability to excel in spelling, but because she could only believe in the failure, and confirmed it each time she was tested, it was so; she owned it, lock, stock and barrel. Nancy, now age 50, continues to struggle with spelling because she is allowing herself to be controlled by the experiences of her past--those of a little girl. There is nothing true for you unless you believe it to be so. If you don't take control of your life someone or something (like your past) will.

Now let's discuss your relationship to the past. What kind of memories came to mind as you thought of your past? Look at your picture. Did you remember the terrible and/or less than positive experiences of your past, or did you draw a picture of the loving, wonderful times that shaped your life?

During private counseling sessions in my office, I will often ask clients to tell me about a positive experience from their past. I am still amazed at how many people will automatically tell me they have never had a good time. Although this may be true to them on a conscious level, I highly doubt that out of all the time they have spent on earth they did not have even one pleasant experience. What these people are experiencing is known as "selective thinking," which can occur only at a conscious level. Selective thinking is the ability to *remember to forget*. This is a part of that servo-mechanism, but one that can be unhealthy and sometimes detrimental. If the mind has been programmed to believe that nothing good has ever happened, then the unconscious will hide all the loving and positive information. What you believe *will* come to pass as reality. In this case, only the unhappy and negative experiences can be found.

Perceptual Filters

Imagine that you have just put on a pair of blue tinted sunglasses. If you were to look around the room and describe what you see, everything would appear blue to you. Therefore, everything you describe would be incorrect because you are filtering it all through the color blue. You can think of your perceptual filters as blue-tinted glasses, which would represent the experiences of your past (either negative or positive). Your perception is like a filtering system. The secret to awakening your genius is to know that you can choose to take off those blue tinted glasses at any time. If a blue world no longer works for you, change the lenses to green, or red, or better yet, wear no glasses at all.

What about your perception of the past--is it truly yours, or is it your father's or mother's or a friend's opinion of your past? Look at your picture again. If your perception of the past is "bad," then you are denying your mind the ability to remember what is good.

The ability to *remember to forget* will become a part of every aspect of your life, including your classroom experiences and other learning activities. What you learn today will become tomorrow's past. If your perception is that the past is bad, then when you get to the test, which is all about yesterday's information, you have started out from a negative frame of reference.

Let's look at it another way. What if your perception of test taking changed. Suddenly you think of test time as fun and exciting. For you, tests are a way to discover just how well you learned and retained information. Now your whole perception has changed--it is seen through the filter of fun!

Which of these two perceptual filters will give you better results? Clearly, the test taken through the perception of a miserable past would deem results clouded by negativity. How about with the perception of fun? Well, you know how you can perform any task well when you are having fun.

Step up now and own your past--the good, the bad and even what falls in between. Own who you are, what you have been, and, most importantly, who you are becoming. Only then can you transform the present you into a powerfully positive person.

YOUR MEMORY

Drawing your past is also a key to how your memory works. We tend to remember life's odd and unusual events. You won't hold a long-term memory of your everyday drive to work or school, but you will remember if there was an accident or if something strange or eventful occurred.

Remember the holographic mind from Chapter One? Because the holographic mind stores information with emotion, this is also where your memories are formed. You hear a love song on the radio and your mind shifts from the lyrics you are hearing to a memory, perhaps of a past love, or a special event that is triggered by that particular song. The hologram of the brain experienced an auditory trigger.

You are out in the woods and smell the scent of pine; instantly you are transported back in time to a camping trip with your father. The forest surrounds you, the campfire is burning low, and you and Dad experience a moment of perfect quiet. This is a holographic experience triggered by the olfactory (smell) sense.

Try this one on. You are together with your family members at a picnic. Suddenly you begin to recall memories that had long ago been stored away in the closet of your mind, but because of the stimulus of the family gathering, perhaps with aunts, uncles, or cousins whom you have not seen for awhile, the holographic brain triggered those long-forgotten memories.

Your thoughts, memories and beliefs are pieces of this hologram that make up who you are and these qualities are also the part of you that can place limits on what you can become. Awakened geniuses know how to access the holographic memory through internal means. They build a perfect memory and recall system. As you read through this book and practice the Self-Help Dialogues, you will be training your holographic brain to bring you the memory of a genius.

YOU ARE INCREDIBLE!

You, right this very moment, are incredible! You exist as your physical body, your emotions and your mind. But, you also are made up of much more than this. Somehow, right this very moment, there is a part of you building your body. It is not only building it, but is also controlling every function. Your breathing, heart rate and digestion are only a few of the array of bodily functions maintained by this incredible part of you.

All of your past memories and habits are also being held in place by this power that works outside of your awareness. It is a force greater than your emotions and one to which your conscious mind is simply incomparable. Therefore, in this training we will consider it your *other-than-conscious* mind. This is the greater part of you, the part of which you are unaware. It is your vast potential that, up to this moment, has been untapped.

THE PHYSICAL BRAIN

It is a common belief that the human brain is somehow vacant at birth, and as the child begins to grow and receive stimulus the neurons start making connections. Science is now discovering, however, that the reverse may be true. The infant is actually born with many *more* neuron connections than most elderly adults. It may be that learning does not happen by making neuron connections, but rather by "weeding out" those that are not used. If this is true, then we are all, literally, born with brilliant minds and we must *"use it or lose it."*

Most babies will, in the first weeks of life, babble almost every possible human sound. Yet, these children will later lose the ability to make the sounds that are not a part of the language which they have been trained to speak. Therefore, the child's environment and the thoughts and ideas presented will play a tremendous role in brain development.

Scientists claim that in our society the average individual uses only 5% to 10% of the brain's potential. Imagine what your life could be if you were able to stimulate your mind into activating even a small portion of that unused potential!

Begin by asking yourself:

(I strongly recommend that you take your time answering these questions and write your answers down. This will give you a frame of reference for your Self-Help dialogues.)

1. ***If I could have anything at all, what would it be?***

2. ***What would I need to do to attain it?***

3. ***What stops me from doing those things now?***

4. ***How can I overcome those challenges to attain the life I desire?***

These questions move us into the realms of *imagination* and *positive thinking*--the qualities that will allow you to awaken that other 90% and keep those neurons firing! You might now say to yourself, "I already have a positive attitude," or, perhaps "that positive stuff doesn't work; all I need is willpower!" The following demonstration will take only a second to prove to you that the imagination and positive thinking are far more powerful than "willpower" can ever be.

WILLPOWER VS. IMAGINATION

Read through the following words gathering together all of your "willpower" ...now **DON'T** THINK OF A RED FIRETRUCK.

What happens? What most people find is that they automatically thought of a RED FIRETRUCK. This is because when the "will" and the "imagination" are in conflict, the imagination always wins. Most people spend a great deal of time telling themselves what they don't want instead of what they <u>do</u> want. Even with every ounce of willpower they can muster, they somehow end up with more of the same old thing. This can be explained very simply. The mind works in pictures, not words. Therefore, it's not what you say to yourself that's important, it is what you imagine.

From this idea came a profound analogy -- IMAGE-a-NATION -- or as the term has come to be known --*imagination.*

When looking at a map you cannot possibly see the territory--it can only be an *"image"* of the *"nation."* Each one of us has an *inner nation* that exists only in our heads. Many people believe that simply because a thought exists in their head it must naturally be present in the mind of every other person as well. Obviously this is not the case.

It is true that along life's path we have collectively agreed on some ideas and given them names, like the *sky is blue* and *water is wet.* However, there is no way to prove that either of these statements is true because this is just the information that is filtered through our senses.

As I stated earlier, our mind is likened to a computer; before it can process any new information, it must first access what I call *resident memory.* This resident memory is the space provided for you to make recollections of all past experiences, thoughts and actions and relate them to present experiences. Resident memory is what makes sense of your day-to-day discoveries. You can think of resident memory as the filters through which your holographic mind will perceive the world and your experiences.

Let's say you have just run into an old high school buddy on the street. This was a friend with whom you have shared some of the best times of your life. The memories involving this person are naturally stored with a pleasant and positive context. Therefore, regardless of the current conversation's content, the resident memory of the joyful experiences will influence the communication. You will probably express excitement and pleasure in seeing your old friend. Your conversation will most likely be quite animated, your eyes will show pleasure and your mouth will frequently curl into a smile. You are greeting this person through the filter of a positive memory.

On the other hand, if you meet someone with whom you have had unhappy experiences (or even if a new acquaintance has an appearance similar to someone who crossed you in the past), the resident memory of dislike will come up. It will not matter what that person says or does, the current information will be filtered through negative memories. Therefore, it is unlikely that this person will be able to do anything right by you. Old, unhappy memories are *residing* in your *memory*, thus the term, "resident memory."

Most happy people have resident memory programs that support positive feelings and activities so they can see life like a child. They are *imaging a nation* filled with possibilities. Most unhappy people are running resident memory programs of past failures, anger at other people, or guilt over some past event. They are *imaging a nation* filled with failure. Abraham Lincoln tells us, *"People are about as happy as they make up their minds to be."*

Each one of us has the potential of a genius. Remember the old cliché, *"If at first you don't succeed, try, try again."* A failure isn't the end, unless it means that you stop pursuing success. In the words of the founders of Neuro-Linguistic Programming, *"There is no failure–only feedback."* Successful geniuses have already figured this one out. What would have happened if Alexander Graham Bell had listened to the people around him who said he was crazy to think he could send a voice across wire? Our generation now takes his invention for granted. But do you realize how far he had to stretch his imagination to even conceive of such a thing?

Did you know that every computer, from the largest, most advanced in the world, right down to your little desktop, performs every function solely with the numbers **one** and **zero** (or **plus** and **minus**)? If some genius somewhere had not stretched his or her imagination to conceive the possibility of a number zero, our civilization would have been rendered incapable of technological advancement. Zero never meant a whole lot to you or me; we take it for granted as a part of the math we learned in grade school. But to the geniuses of an earlier time, it was the dawn of a whole new world. It's hard for us to imagine, but the rules for that time were that numbers consisted of 1, 2, 3, 4 ...up to 9. It took sheer unadulterated genius to conceive of the number zero; someone zealous enough to tap into his or her creative spark and willing to think outside of known rules. Think about it: one or more zeros, placed on the end of each of those numbers, and you have ...well, infinity!

It's uncertain as to who should receive credit for such limitless thinking. It has been speculated that the Hindu civilization used zero 10,000 years before the birth of Christ. Its evidence was surely found in the Babylonian time period, 250 BC, where it appeared in their place value system. Unfortunately Diophantus of Alexandria didn't know about these ancient discoveries, for in 210 AD he accidentally "discovered" negative numbers, but he considered this possibility totally absurd and discarded his entire project. Today we take zero and negative numbers for granted as a simple process of mathematics.

Sometimes, in finding solutions to your own problems you may feel, like Diophantus, that a particular solution is totally absurd. "No way," you might say, "This can't possibly be true!" Think of what Diophantus missed with his limited thinking. Perhaps *you* will want to stretch your imagination, look at the idea from all angles, uncover all possibilities, and then decide what will be true for you.

Imagination is the secret ingredient in life itself. If people imagine themselves "rich," regardless of the amount of money they have, chances are they will remain healthy and strong and live an abundant life. If a rich man imagines himself poor or on the brink of losing his fortune, it will be likely that trouble and sickness with follow him through his life. All that we are searching for in life is always present within ourselves. This is something like Dorothy and poor little Toto trekking all over Oz in search of her dreams, only to discover that what she was seeking had been right in her own backyard all along.

Nothing is outside of you. The sooner you come to believe in the power of your own imagination, the easier your life will become. I challenge you to stretch your mind to the possibility of accomplishment *beyond* your goals. What if, in an attempt to accomplish your goals, you awaken within you a latent genius with the ability to overcome one of the world's pending shortcomings? What would you feel like if you found the cure for AIDs or cancer? Think of the people who would benefit from your achievement. And to think, you did all of that just by helping *yourself*! The reality is that until everyone in the world imagines a world of health and vitality for all, there will always be disease and woe. Words are powerful, but remember, *a picture is worth a thousand words.*

WHAT THE MIND DWELLS UPON IT MUST BECOME!

This knowledge is your most powerful tool for changing your world or enhancing it, because it is where we all must live. We each have an inner world and an outer world and they usually don't match. The main reason for this is that we all delete, change and distort reality to fit into our *mold* of what reality is. Not what really *IS.*

We don't have any way to know what really is; we can only see, hear and experience life through our perceptual filters. These filters are what screen out reality and they are called the senses. I once read a story about how the Indians responded when Cortez landed his boat in the new world. The author claimed that the Indians could not even see the ships in the harbor because their minds were incapable of conceptualizing a ship. Similarly, it had once been reported that when the Indians saw men riding on horseback they believed that man and horse where one and the same and could not be separated. They thought they had encountered gods that were half man and half animal. The Indians simply had no mental conception for the possibility that a man could ride on the back of an animal. They could only presume they had come in contact with a god.

Are you perhaps one of those people who is afraid to stretch your perception of reality? What if you could believe that your brain has already, instantly, scanned this page several hundred times? You already know everything written here and are simply reading it again because you haven't yet grasped the reality that you are truly that intelligent. Did you know that every human being has a perfect photographic memory? It's true. Unfortunately, photographic memory lasts for about one-tenth of a second, and then almost all of the information is consciously forgotten. Young children have photographic memories that last much longer. For some reason, however, the ability is lost as children grow older. Could this be attributed to lack of use? Do the neurons that generate photographic memory stop firing when the skill is no longer serving a purpose? Anthropologists tell us that in cultures where people do not learn to read and write, persisting photographic memory will continue throughout adulthood. If this is true, is our educational system actually creating limits and barriers to our genius potential? Perhaps our kindergarten curriculum should consist of photographic memory classes and leave the ABC's for later when the child can learn them at a glance.

Please don't think that because you are able to read this page your memory skills are now lost. Remember, everyone has a *perfect* memory. This is true whether you are one or ninety-one. Your mind takes in all information and remembers, either consciously or unconsciously, everything you experience through the five senses. It is the process of *retention and recall* that needs improvement and here, again, we are within the realm of your belief system. If you truly believe that you have a perfect memory and that you already have a faultless recall system, the brain, being a servo-mechanism (goal oriented), would continue to modify your skills until this becomes a truth. This is the purpose in mental imagery (our Self-Help Dialogues) and affirmations. If the statements are believable to you, then they will come to pass as reality. If not, however, they are just words.

WHAT IF?

Remember Nancy, our horrible speller? She had herself so stuck into the conscious belief that spelling was a miserable chore, she found it impossible to think outside of those pre-set boundaries. Her conscious belief created the limitation and then she *was* terrible at spelling. In the science of Neuro-Linguistic Programming (NLP) they developed what came to be known as the "*What If?*" Frame. The word "Frame," meaning frame of reference. Think of anything that you would like to do or become and ask yourself the question "*What if?*" As an example: *What if I scored 100% on my next test? What would I see? What would I hear and feel? What if it just happened?*

The "*What-If*" Frame gets your mind focusing on possibilities and opportunities--because *choice is always better than no choice.* Without possibility the mind is stuck. It is like an idle engine with no gears to start the movement. When one starts to think "*what if*," one moves from a stuck state into a natural flowing state--which is where solutions reside. The "*what if*" frame will bridge the gap between the right creative mind and the left logical mind. Even 2,000 years ago they knew that faith without works was useless. The "*What If?*" Frame takes your mind from "*I hope,*" or "*I wish,*" to ***"I can!"*** and ***"I know how!"***

I once counseled an athlete who was a high school basketball star. "Pete" was a tall, handsome young man with sandy hair, green eyes and a gleaming, broad smile. (I didn't actually see the smile until later.) He had a mature air about him and could easily have been mistaken for a college student. It was Pete's father who had made the initial appointment with me; therefore, Pete walked into my office with chin on his chest and smoldering eyes. I was soon to discover that Pete had only one interest--basketball. When his father brought him to see me, it was because his grades were so far below average that he was no longer allowed to play with the team.

I started by asking Pete, "What if you could have the same type of enthusiasm that you have on the basketball court, but in your math class?" I don't know what Pete was expecting of me, but he simply gazed at me with the most interesting puzzled expression. He didn't know how to respond--he had never even considered that math could be fun. As I explained it to him further, I helped him to understand that math may never hold the same thrill as making a three-point shot, but, in his mind, success on a test in math could become equivalent to scoring points on the basketball court.

Because Pete's mind was now open to possibility, our conversation developed with a flow and sense of direction. We discussed his other subjects and how he might begin to think about them differently as well. I helped him build his sense of curiosity. "*What if*" he could walk into the classroom and establish the same kind of curiosity that he had when first learning how to play basketball? He was not to think of the way he was playing now that he had perfected his skills, but before that, when he was developing a series of learning techniques. For example, the way in which he watched other players, the way his mind focused on their form and moves, or even his own daydreams of making the perfect lay-up. The attention that he gave on the basketball court was probably far superior to any attention he had ever given a teacher.

Pete left that first session with an agreement that when he entered a classroom he would begin to think with the *"What If"* Frame. *What if* his teacher was not only talking about math, but giving him ways to put mathematical sequences together so they could be just as exciting as dribbling down the court with ease or executing the perfect pass?

In that first week, Pete began to notice that his time in the classroom could be much more exciting than he had ever thought possible. His memory and retention of information was enhanced simply because he had a new frame of reference for learning. When he came back for his follow-up session, I explained to him that the information he had stored with curiosity was vivid in the mind and could be easily recalled. As Pete learned to store information from a curious mind and with excitement, his memory was greatly improved. Pete found his grades slowly improving and soon he was back on the team, enjoying the limelight *and* preparing himself for a college education.

SELF-DISCOVERY: WHAT IF?

What if you could be in the future looking back over time? From that future place, you could obtain all the information you would need for accomplishing your goals. You could see, feel and experience everything that you did right to make your success a reality.

What if you felt so relaxed and comfortable with learning new information that all the time you spend studying is instantly applied to your testing situation?

What if, during every test, your mind and body agreed that the information can flow from your *other-than-conscious* mind into your conscious awareness?

What if it was fun for you to succeed at everything you attempted?

What if you could not only pass the test without much conscious effort, but you *expected* to pass each test?

Doesn't it feel great to be relaxed and comfortable with what you are learning? Isn't it fun to succeed? Wouldn't it be great to pass all of your tests without much conscious effort? Again, your mind works for you; the "*What If?*" Frame helps you to focus on what you want and to build the belief that you can have it!

If you don't take the time to show the results on the magic screen of your mind, then your mind thinks it is doing exactly what you want.

EMOTIONS

Remember the *Incredible You*--that awesome combination of mind, body and emotions? Your emotions, or what some would term your state of mind or your physiology, also play an important role in what you perceive as reality and who you believe yourself to be.

Can you remember a day when you felt "on top of the world," feeling good about yourself and the world around you? If something went wrong on that particular day you could easily handle it. And then along comes another day when you're feeling less than positive about yourself and the smallest of things can "topple the apple cart." This is because most of our actions are simply responses that have been programmed into us since birth. Most people no longer live life, they simply survive it. They make it from paycheck to paycheck, looking forward to Friday and dreading Monday. When we feel dissatisfied with life, it's time to realize that we have been living a life of responses to our perception of reality, not reality itself. Without this recognition we are sure to stay on a crash course to self-sabotage. Life in itself has no inherent meaning. We, as people, give it meaning through our emotions and in relation to our past experiences.

All actions have equal and opposite reactions. As humans we have learned how to react to life through our parents, family members and friends and by our mistakes and successes. In the following section you will discover how to change the past programming so you can begin to live in the moment, at the point of choice, where you can create the life you desire. Your mind works in sequence. You move from one event to the next, responding to every stimulus that you encounter; all that you see, touch, taste, smell, hear, feel and do. Each thought and every action is going to bring you to your next experience.

SELF-DISCOVERY:

I challenge you to stand up while reading this next part of the text. Go ahead now, stand up! Don't worry, I'll let you know when you can sit down again. What have you got to lose but a few bad habits!

While standing I want you to think of a circle in front of you. Use your imagination to give this circle a color and make it your favorite color. When you have a color in mind, imagine that in your circle there is a new skill or ability that's just for you. It could be concentration ...It could be assertiveness ... It could be the feelings of relaxation ... It could be the feelings of happiness ... It could be all those things and more. Use your imagination to the best of your ability ...create it and then step into it. When you step into it close your eyes and imagine that the color is filling you up ... as if your body is an empty glass container that could be filled with color. I'll wait right here, until you have filled yourself with that color and are ready to continue reading . . .

You have now filled yourself with all of those wonderful emotions ... those feelings of completion and satisfaction and much more. In fact, you will find that when you are done with this challenge and because you have read these words, your mind has already begun a process of creating this Circle of Power out in your life... so that your circle of power will work for you whenever you may want or need it.

Think of a time and a place when you would like to simply step into a space of personal power. It could be when you get home from work or school. It could be in front of the refrigerator. It could be when you open your math book or start your English assignment. It could be when you are in the classroom.

Guess what? ... Your mind has already thought of three places ... hasn't it? Think for a moment of three places that all you would need to do is step into that circle ... the one in front of you ... and you could step into your space and place of personal power. Practice this process:

Step One: **Think of a negative emotion.** *(make this an emotion you would like to change, allow it only a moment, and then move on to the next step)*
Step Two: **Step into the Circle of Power.** *(mentally imagine it filling you up)*
Step Three: **Roll your shoulders back, lift your head up** and allow your imagination to take over. *(breathe the way you breathe when you feel happy and proud)*
Step Four: **Imagine what you would need to change so that you could enter into the experience with the thoughts of possibility.** *(Open your mind to the possibility that there is another possibility. You are building flexibility. You may want to use the "What If?" Frame here.)*

Okay, you can sit down now.

We all have a Circle of Power that we carry around with us. If you stop for a moment and think about what you do well, what you are confident about, whether it's a sport, reading a book or talking to your friends, where you can be totally who you are in the power of the moment, that is your Circle of Power.

You need only think of those times in your life when you lack confidence or think of yourself as less than you want to be to understand how you might benefit from the Circle of Power. In truth, you are in control; you have control of the five senses, the way you see, hear and experience your life around you. The Circle of Power puts you into the best situation for processing the information around you.

Some students use the Circle of Power for class speaking assignments. Right before getting up, they imagine their Circle of

Power so they will feel confident, will possess all needed communication skills, and will allow the thoughts to flow from their mind to their mouth. Some students imagine taking the Circle of Power out of their pocket and throwing it on the floor so they can step into it. Some students use it under their desk for whenever they are taking a test. Some students use it in sports: while shooting a free throw, when up to bat, or just before stepping onto the track for a race.

Whether you are experiencing tests in school or tests in life, the Circle of Power is an effective and easy way to give yourself an emotional lift whenever you need it. Many sales professionals have learned to use the Circle of Power before making a sales call, greeting a customer or lifting the receiver of the telephone. Even the day-to-day encounters of life, such as discussions with family or friends, can be positively influenced with the Circle of Power.

The Circle of Power has an infinite number of uses, and only you know just where or when you will need it. Some people even place it right at their door just before leaving for school or work in the morning. They can begin to use it from the moment they step out and start their day. You might remember your kindergarten teacher telling you to "put on your thinking cap," for the day. She probably didn't realize what she was saying, but she was actually helping you to change your state of mind to one of learning--you were creating a "circle of power" right on top of your head.

WHERE ARE YOU GOING?

Awakened Geniuses know exactly where they are going; they know what they want to achieve. Perhaps they are not yet aware of *how*, specifically, they will get there, but they know the direction in which they are headed. The ability to *improvise* is a very important ingredient of the genius mind. A genius learns early in life that there are no victims. They have chosen their place in the great scope of life and if they don't like where they are, then it is up to them to change. Geniuses understand that they are in full control of one universe--it weighs about three pounds and exists inside of their own head. Geniuses also know that there is a source and power far greater than their little human form that controls and shapes the destiny of the human condition--and it resides in their *other-than-conscious* mind.

If given the opportunity, would you change any portion of your past? If you said "yes" to changing even one small segment of your past, then you have given a direct suggestion to your *other-than-conscious* mind to go back through your memory banks and clean up old behaviors or perhaps clear away a limited belief. A genius knows that the greatest control comes from taking command of one's own mind and that the *control is in letting go.* I know that this seems like a contradiction in terms, but it is completely true.

Whether you are attempting to control another person or a situation, it is just when you are trying the hardest to control that you are most out of control. I think Buddha said it best, *"He who angers you, conquers you."* When you begin to spit and sputter your fury all over the place, it may feel as if you are in control of that moment ...but are you? The genius knows better than to be affected by another person's opinions or actions. True geniuses know what is "true" for themselves and can allow other people to have their own truths. The reality is in the moment. So the more spontaneous and positive a person is, the better the world seems to work. Some folks are under the impression that things have to be going "right" for them to feel good. This is the attitude of a loser. Anyone can feel good when things are going right; it takes a genius to create a perpetual state of excitement about life itself.

What would happen if you could look at a problem with new eyes--the eyes of wonder? And you said to yourself, I wonder how great things will be when I solve this problem and get on with my life? And then you proceed to solve the problem with the attitude that the solution is at hand, even if you have no idea what it is. The real "brains" in this world know that our minds store solutions with problems and problems with solutions. As an awakening genius, your task is to focus on the solution and simply use the problem as a stepping stone to an even greater success.

ARE YOU SETTING YOUR OWN LIMITS?

I was once called upon to work with a client in Scottsdale, Arizona. Scottsdale is a lovely desert community just outside of Phoenix. Its long boulevards and graceful stucco homes are surrounded by an array of desert foliage. Saguaro cacti and flowering desert ferns abound and are surrounded by rocks and pebbles of all shapes and colors. I thought I knew what to expect when making a call in Scottsdale.

However, when I pulled into the drive of this particular client's home, I was surprised to be greeted by the unending expanse of a swirling wrought-iron gate. An elaborately grated stand stood at salute in the center of the drive. This, I soon discovered, contained a blatantly high-tech telephone system. I picked up the receiver and an aloof voice immediately asked for my name and the purpose of my call. I noticed myself growing self-conscious as I became aware of the camera lens hidden in the bushes and directed at my modest Oldsmobile.

In a moment the gate began its slow opening grind. As my car advanced to the stucco and stone masterpiece that sprawled before me, I felt myself draw in a deep breath ...I was in awe. Somehow I couldn't imagine anyone calling this monstrosity "home."

I should not have been surprised to find another sophisticated-looking telephone awaiting my arrival at the grand entrance. When I picked up the receiver this time, the same haughty voice granted me one word ..."Enter." I heard a brief buzz and the massive doors clicked their allowance for me to push them open and step through. I had the feeling that I was entering a fortress. Surely this person was born into money, was my first thought; there is no way anyone could attain all of this in one lifetime.

My work with this client, I'll call him "Frank," was loaded with surprises. He readily confided to me that he had, in fact, been through bankruptcy three times. This was his fourth go as a millionaire! He carelessly shrugged his shoulders as he described his difficulties in making the first million. The rest, he said, was really quite easy. Clearly, this man has an entirely different perception of a millionaire than the average person.

Frank had been born into a large Depression-era family. He described how his family had stood in line for hours hoping for a loaf of bread and some watered-down soup. As a young boy he made up his mind that one day he would make a million dollars and had convinced himself that the million would somehow make up for all of the pain and loss of his youth.

I found myself entranced by the gleam in Frank's eyes as he described the obstacles he had overcome in attaining that first million. His manner became animated, almost exuberant, as he told me of his many escapades as an up-and-coming businessman and of the lessons he had learned the "hard way." By the age of 40

Frank was a complete success; he had made a million dollars. Yet it was just about that same time that his life began to fall apart.

Frank, his body visibly deflated and his eyes dark and somber, described the years that followed. His marriage fell apart, he made some unwise investment decisions and allowed himself to be taken by a "con man." Frank was broke and in bankruptcy court before he knew what hit him.

With those experiences behind him, Frank gained a new resolve. He would never lose his million again! With an entirely new business concept, he set about rebuilding his fortune. Indeed, he made another million and promptly lost it all again. What happened? He had asked himself this question endlessly, yet no answer was forthcoming.

I explained to Frank that he had done a great job of programming success, but he had set limitations on his success. Once he made a million dollars, he no longer knew what to do with himself. He promptly lost the million so that he could go about achieving the goal he had programmed--making a million dollars. He had called me in desperation as his fourth empire was on the verge of toppling.

Although I was no financial advisor, I could tell Frank was setting up a self-fulfilling prophecy of failing one more time--which he had effectively trained himself to do. He would begin a process of drinking, making poor business decisions and wreaking general havoc upon the dynasty he had so painstakingly built.

What I had to help Frank understand was that not only had he programmed his million, but he had also programmed his fortune's demise. Without his awareness, he had placed a limit on himself. He could indeed work like a trooper to make a million dollars, but then he would promptly need to lose it, for his self-confidence and self-worth were not at the level that could sustain that type of income. He believed in the struggle, but not in the success.

Frank began to realize that money itself is not an appropriate goal, and he would need to develop a more well-rounded attitude about his finances and his life. Now that he had made his fourth million, there was no longer a need to be so risky; he could relax and enjoy the fruits of his labor.

He wanted a more conservative approach than he had practiced in the past, and I told him that it would come as he learned to live his life one day at a time. As Frank and I worked together over the next several months, his one-day-at-a-time attitude showed signs of blossoming.

Years later I received a surprise call from Frank. No, he didn't need my services any longer; he just wanted to call and thank me and let me know that he still practiced the processes of relaxation that I had taught him. Each day during his "my time," he could come to grips with his day-to-day life and stresses and then plan his next adventure. Indeed, Frank's life had now become a series of successes as his fortune had multiplied many times over.

It is my hope that everyone who reads this book will expand their awareness to a true understanding that money in and of itself is not an appropriate goal. It is the emotional state, the feeling of well-being, that our human nature truly seeks. There are poor people who are truly happy--blessed with the gifts that the world has to offer. There are wealthy people who have more sadness, grief and stress than anyone I have ever met.

Make a promise to yourself today that if you ever find yourself feeling depressed, anxious, angry, fearful, or holding resentment toward another human being, you will immediately take that limited belief and place it behind you. When that space in front of you becomes clear, you can begin to look forward to even greater possibilities. The best goals in life are those that not only support you as a total being, but support the world as well. When you are of service to yourself and humanity, you naturally feel good, you feel alive, regardless of your financial situation or station in life.

"Your mind is the most valuable thing in your possession. It would cost more than twenty billion dollars to prepare a sensitive instrument which could function like the mind; still it would not be able to register and broadcast feelings of love and other higher emotions of devotion, reverence, compassion, sympathy, etc."

Yogi Gupta

CHAPTER FOUR

INCREDIBLE YOU!

Now back to the incredible YOU--and your mind. The brain has often been likened to a computer and it's true that there are many similarities. However, there are some unique and profound differences. The brain is made up of nerve fibers (neurons) that are alive and can grow. This alone makes the brain infinitely more intelligent than any modern-day computer. These neurons are constantly firing synaptic connections with other neurons and processing the information at an incredible pace. Yet, just like a computer, the brain must first have a base operational system from which to relate the incoming information. In this way, your mind is much like a computer, one that is self-aware and, well, almost predictable!

Predictable because your operating system has been programmed all through your life and by a variety of sources, such as your experiences, your internal thoughts, and everyone else to whom you have access.

SELF-TALK

Researchers of the mind have found that in each second as many as 20,000 bits of information enter the mind through the senses. Along with all of that sensory information, we are also talking to ourselves at an incredible rate of 5,000 words per day or more. This is your *internal dialogue* -- the part of your mind that is always talking to you. These are internal conversations that take place constantly and can either motivate or sabotage success. How can we possibly control that continual flow of information? We can't--our unconscious or *other-than-conscious* mind does the controlling for us. Just how this occurs will become clear to you as we discuss how the brain and brain-wave activities work. A genius becomes aware of his or her own *internal dialogue* and begins to change the messages that are no longer creating success.

It is my hope that by now you are beginning to understand the extraordinary potential (genius) of your *other-than-conscious* mind. As you consider the functioning of your own internal dialogue, you can understand how those internal words can make a difference in your life--either positive or negative. How is your internal dialogue controlling you? Have you ever talked yourself into or out of something you truly wanted to do? Perhaps you had planned on spending an evening studying for a test and ended up watching TV instead? What about that time you planned to exercise after work and ended up going out to dinner with friends? How often do you talk yourself into or out of what you truly want to accomplish?

We will be learning in this text how each word of your internal dialogue creates pictures in your mind and how those images help you to relate with your world in a positive or negative way.

It is everyone's right to get what they want. Yet, many pessimists will claim that extenuating circumstances stand in the way of success. What they don't understand is that these outside influences are all a part of their beliefs and values. A genius is one who creates ongoing "circumstances" that will in turn produce the probability of success no matter what the odds. History books are filled with stories about people who made it from the gutter to the penthouse or from convict to social reformer. What did these people have that seems so elusive to others? They were somehow able to make a quantum leap, from blaming their situation on some external source to taking responsibility for their life. In almost every "rags to riches" story there was a moment of reckoning; the realization that each individual is "at cause" in their own life. Once that mental choice to think differently was made, they became interactive with life so that when opportunity knocked, it was heard. Sometimes we think we are open to all possibilities, but because of preset beliefs we limit our choices for success. What we think we want and what we are focusing our attention on are often two different things.

Let's use a metaphor to help you better understand the laws of success for making a change in your life. Let's say that you work like an airplane. Aviation experts are fully aware that from the moment an airplane takes off in one city, and until it reaches its destination in another, it will be off course for about 90% of the trip. Yet these pilots are always confident they will reach their target and, with agreeable conditions, right on time. This is because the pilot and co-pilot know exactly where they are headed and have a map to guide them when they move off course.

How many people believe they are doing exactly what it takes to bring about success, yet are totally unaware of when they are moving further and further off course simply because they never took the time to review their *course of action*? Some people, even when informed that they are moving in a direction counterproductive to success, will deny it and continue to go against the proven laws of success. Can you imagine that?

Back to our airplane. Did you know that when an airplane takes off, it uses 110% of its horsepower just to get off the ground, but once in the air it takes only 40% or less to keep it airborne? That's right, during take off most airplanes must use engine power beyond their "capacity" in order to get off the ground. How many people do you know who will not even start a project because it involves too much work? They probably have only thought of that initial 110% necessary to get the project off the ground. Geniuses know that once a task is started, it will develop a life of its own. All they will need to do is continue nurturing their dream one day at a time and maintain their physical involvement in getting the job done.

What does this have to do with you? Well, let's start with the facts. It's okay to realize that when you begin to change habits and patterns that have been with you for a lifetime, it may be difficult at first. Once the sequence of change has occurred, however, it becomes easier and easier. Soon the new, more appropriate behavior is dominant and the old patterns melt away completely, once and for all.

Still using the airplane concept, let's say you want to change an unwanted habit or behavior. Then, let's just say you put all of your effort, even more than you consciously know about (110%), toward the change; would it not be safe to say that the new behavior would come into being? And, after some time of practicing the new behavior, wouldn't it be safe to assume that it will grow easier and more natural for you? Now, wouldn't it seem highly probable that you could start to use that same consciousness to think of other choices and possibilities, perhaps even other positive changes that you could make? And, since you've done it once, wouldn't it seem obvious that you could make the next change with even less effort?

OUTCOMES

The first step in making changes in your life is to know what you really want. We will use the term "outcome" to represent that which you want to achieve--it means the successful conclusion that you desire, or what *comes out* of your efforts.

My first experience with "outcomes," came when I was in high school. At that time my father was actively involved in controlling alcoholism, and he relayed much of what he was learning to his kids. During the summer between my freshman and sophomore year, I had gotten myself into just enough trouble to get me grounded but good. I was confined to the basement with nothing to do and could see no one. Dad knew that grounding me never seemed to do much good and that was what gave him an idea. On the second day of my confinement, he brought me a tiny book with, "As A Man Thinketh," etched across the cover. He informed me that I was to read it, cover to cover, each day. I shrugged my shoulders, took the book from his hand, and dropped down on the bed. Might as well get started, I figured, since there was nothing else to do anyway.

It's hard for me to describe what I experienced next. As I began to read, it was as if something dark and painful that had been lurking within me was suddenly fading away. I was opening up to a new world of hope and possibility, one of freedom where I was in control of my own life. The thoughts which those little pages invoked left me feeling light, exuberant even! It was all so simple. All I had to do was set my mind to what I wanted to achieve, imagine that I had already done it, and then do it. How could something that had seemed so complicated now appear so simple?

I found it easy to follow Dad's instructions. Reading those pages diligently each day, I absorbed more and more of its truth each time. Finally, I decided on a specific goal and felt ready to take action. I eagerly approached my father, who was busy reading the evening edition. I tried unsuccessfully to hold back my enthusiasm. "I have set a goal," I announced boldly. "I saw myself achieving it, and am now ready to begin."

"So what's your goal?" he asked, peering at me over his reading glasses.

"I'm going to be captain of the football team!"

I had announced this with such aplomb that at first my father was at a complete loss for words.

"Now Patrick," he said in his most matter-of-fact voice, "You have to set *realistic* goals."

If you could have seen me through Dad's eyes, you would probably have said exactly the same thing. I was not just short; I was 4 feet 11 inches of solid skin and bone! I had always been the shortest in my class and was equally as scrawny. I remained undaunted, however, certain that becoming football captain was not only possible, but would indeed come true.

I continued to read "As A Man Thinketh" every day, even after my Dad had completely forgotten that I was grounded. A few weeks later I was offered a job at a company that manufactured mortuary block. The pay was terrible and the work was grueling, but I took it anyway. Every evening I would allow my bicycle to slowly coast me home to a hot bath and bed.

But then, something began to happen. I noticed myself changing. All of the strenuous work gave me a voracious appetite. No amount of food could satisfy me. I began to discover muscles I never before knew existed. I felt myself growing stronger and more graceful every day. Lifting those huge bags of cement and heavy blocks didn't seem so difficult now. And, best of all, I grew--four inches that summer.

On that first day of football practice my battery was on high charge. I ran up to the coach to let him know I was ready to play any position.

"Who are you?" he asked, squinting as he looked me over. "You that new kid from Chicago?"

"No coach," I answered, "I'm Patrick Porter; don't you remember me?"

"Porter? Hmm." He peered closer. "Mike's brother?" I nodded. "Well, you've certainly grown up." He smiled. "Let's see what you can do."

In that moment I knew that my days spent dreaming of success were over. It was now time for me to take action--to do what it would take to make my dream a reality. I practiced football every day, rain or shine. I began lifting weights and every morning I went out running to build my endurance. And, to build opportunity, I always chose a route that took me by the coach's house. I not only played football, but ran track and wrestled as well. After practices, when the rest of the team was exhausted and ready to shower, I was headed for the weight room.

My school work also became a priority. I had to ensure that my grades would allow me a place on the team. With my new-found motivation I found that I was actually enjoying school. Along with my brother Mike, I began to research what would give me endurance and make me stronger. I changed my diet, eliminating all junk foods and replacing them with healthy, natural foods.

By my junior year I was one of the hottest athletes at our school. My name was becoming a regular item in the local newspaper and several underclassmen began to approach me wanting to know what I had done to become such a great athlete. While many of the other players regularly shunned the younger set, I started a weight-lifting club and invited them to join in. It was fun for me to see them learning and accomplishing just as I had, and I discovered a true enjoyment in helping others.

Just before the beginning of my senior year, the coaching staff held a private meeting and decided that all the team players, including the junior varsity, should be allowed to vote for team captains. Although it had not been pre-planned, all my efforts with the underclassmen paid off. I was voted, hands down, team captain not only in football, but also for the wrestling and track teams.

As a three-sport captain, I had not only attained my dream, but had far exceeded even my own expectations. Most importantly, I had developed a blue-print for success that has carried me from the attainment of one goal to the next, and it is the one that I use to this day.

Now it's your turn. Go ahead, get a clear picture in your mind of what you truly desire. *See* it as if you already have it. *Hear* all the sounds around you and let what you are hearing reverberate as the sounds of your successful outcome. Really begin to *feel* what it is like when the change has occurred. It's important to make sure that you personally have power over the outcome and that you take direct responsibility for its completion.

DESIGNING AN OUTCOME

Ask yourself each of the questions listed below. Take your time and always think each answer through to the completion of your goal. Choose only one outcome for right now and focus all of your attention on its accomplishment. You can always come back through this exercise for any other outcomes you may want to achieve. (Hint: You'll find that it becomes easier and more precise each time you work through this process.) I highly recommend that you write your answers down and keep them handy for future reference:

1. **What is an outcome I would like to achieve?**
2. **What first step would I have to take to ensure this outcome will become reality?**
3. **Am I personally responsible for the outcome?**
4. **What stopped me from doing what it takes before?**
5. **If I reach my goal, what positive result would occur in my life?**

ACCOMPLISHING AN OUTCOME

The second step is simple. It involves getting out there and *doing* what it takes. Remember the statement that genius is 1% inspiration and 99% perspiration? This example can be taken even further in a personal way. Have you ever watched someone performing a task and thought to yourself, I know a better way to get that done, or that gives me an idea? Perhaps you even came up with a new invention or concept all on your own. I think this has happened to all of us at one time or another. It's that moment of inspiration like a light bulb popping on in your mind. Yet how many times did something stop you from pursuing what inspired you? Unfortunately, this is what happens to most people, and soon they end up seeing someone else fulfill their dream and enjoying the benefits of success. They can only sit and watch it happen.

THERE ARE FOUR TYPES OF PEOPLE:

1. ***Those who let it happen***
2. ***Those who watch it happen***
3. ***Those who make it happen, and***
4. ***Those who wonder what happened***

This text is designed to put you back in control of your life. The happiest people are those who make it happen. In order to really get the outcome you want, need or desire, you must take **action.**

"TO ACT AND TO THINK ARE ONE AND THE SAME"

What stops people from attaining their goals? In most cases it is their **thoughts** and **actions** that stand in the way of success. To entirely accomplish your dreams or goals, you must act as if you already have them and imagine the desired outcome in your mind. Eric Oliver, a business professional who often presents corporate seminars, will frequently tell participants, "You will get what you rehearse; which is not necessarily what you intend." In other words, your *thoughts* are your *intentions* and your *actions* are your everyday *rehearsals*. If you have the goal (*intention*) of owning a brand new sports car, but spend day after day on the sofa watching soap operas (*rehearsal*), is it likely your dream will ever be realized?

Several years ago I was commissioned to develop a tape series for the American Taekwondo Association in Phoenix, Arizona. In my research, I interviewed several karate "masters." Each described a process of mental imagery used by the martial artist to practice appropriate thought, action and response. They mentally imagine the exact move to use for any situation, and they see, feel and experience themselves responding with perfect synchronicity. They can then trust that their *other-than-conscious* mind will take over and do the thinking, acting and responding automatically when needed.

This ability is not limited to martial artists. Just like the black-belt, if you give the *other-than-conscious* mind a goal, and you make it clear and precise through mental rehearsal, then you just need to get out of the way. With the vast resources of your *other-than-conscious*, the details will be worked out for you. It is that *other-than-conscious* that has the ability to access all memories from the past and all future outcomes. From there, your mind will chart a path of action with the least resistance to the attainment of your goals--perfect synchronicity.

When it was time to sit down and put these karate tapes together, I knew it would not be necessary to mention every aspect of karate or even the specific moves or postures. If I simply mentioned a key word or phrase, the martial artist's mind was pre-trained to mentally play out the process, and rather quickly at that. What would have taken two or more hours of physical practice could be accomplished within 30 seconds to two minutes. During mental practice they experience the sensation of perfecting their mind-set and their physical body responds.

After completing the tape series, I asked several Taekwondo students to help me test the results. These students were fully capable of learning and performing the physical motions, but they had often felt awkward or uncomfortable. I asked each student to listen to the tapes daily and to follow the mental rehearsal techniques to the best of their ability. I knew that until the neurology of their bodies actually began to respond, it would be an incomplete process. After only a few weeks of listening to the tapes and mentally rehearsing their moves, every one of these students expressed a marked improvement in their concentration and physical ability. The following exercise will help you to understand how the body and the mind work together.

Discovering Your Neurology

This is probably the easiest of all the exercises. Simply answer the following question: ***Which way do you turn a door knob to open a door?***

If you found yourself reaching out as if to turn the handle, just as if you were standing in front of a door knob, then my point has been made--to act and to think are one and the same or, to think and to act are one and the same. The equation works either way. Even if your hand did not physically reach out, you had to imagine your hand on the door knob in order to know how it turns.

Your body's neurology is also programmed. To get your body in action, you need to get your body and mind acting and thinking in the same direction. Your mind knows exactly how a door knob turns, but you had to make the physical movement to unlock that information. The same is true with unlocking the knowledge of your *other-than-conscious* mind. Your body absolutely needs to be a part of the action.

This may especially hold true for those who seem to have a learning disorder. These people may not have a disorder at all--they simply have a mind that learns in a different way from what society considers normal. I believe that many of the children who are said to be learning disabled are actually just "kinesthetic learners." These people have a brain that transacts information when the neurons are flowing. Most kinesthetic learners always seem to be in motion. They may doodle every time they have a pencil in their hand. It doesn't matter what they are drawing; what's important is that the neurons are firing. For them, when the body is in movement the mind is in action and the information flows from the right to left side, or perhaps vice versa, depending on which is dominant. Kinesthetic learners are capable of learning and retaining the same information as anyone else. Yet they need to store the information in a different way than most people so that it can be unlocked when needed.

SELECTIVE MEMORY

Everyone possesses what is known as *"selective memory."* At some time or another you will select the memories to hold for recollection later. Through your day-to-day life you will generally select and remember those things that are odd and unusual or outside of what you consider *"average"* or *"normal."* Your selective memory can be either your best advisor or your greatest adversary, depending upon how you have trained yourself to use it. Again, the genius understands the natural processes of the mind and uses them to his or her advantage. The secret lies in knowing how to access the simplest procedure to extract the greatest result. Since you are always absorbing the world around you, you will now learn how to access the necessary information, have it available whenever you want it and discard what is harmful or unusable.

We once used a group of seminar participants to conduct a study on selective memory. The room was filled with adults and young adults, and all were shown two separate groupings of pictures to study. Each set of pictures consisted of depictions of popular government officials; the first set included photographs of their true appearance and the other were cartoon characterizations of the same people. Our little research project revealed that the average person had an 80% better retention and recall of the cartoon characters. They remembered the pictures that were odd and unusual, or different from average.

The way the genius *benefits* from selective memory is by using it to heighten internal learning states, thus creating richer experiences (see Optimum Learning State in Chapter 10). By creating an environment for selective memory you are essentially making your learning experiences more memorable. The Circle of Power is an excellent example. Let's say that you are in a lecture given by a particular instructor. As you surround yourself with your Circle of Power, you are creating an experience that is *out of the ordinary*; instantly you have optimized the learning experience. That lecture is now stored with all of the positive thoughts and feelings that your Circle of Power instills. During a testing situation you will simply activate the Circle of Power, and the information will flow into your consciousness. You have selectively set up an *other-than-conscious*-to-conscious response of a perfect memory and recall.

One might also use the selective memory of a positive learning experience, a time when you learned how to do something that you enjoy. (Remember how "Pete," our basketball player, began to learn in the classroom with the same excitement as on the court?) As you learn something new, you can select that same state of *enjoyment*. Suddenly the new learning experience would be different (enjoyable), a little bit odd and unusual. You now can easily select that memory from a state of enjoyment and allow it to filter through. Again, we are setting up the perfect memory and recall.

Everything you ever see, hear or experience is remembered. You have a filtering system (known as your *mind*) that absorbs and assimilates all that occurs around you at a rate of over 20,000 bits of information per second. Those bits of information are selectively stored by your *other-than-conscious* mind; some bits are stored to be remembered and recalled, others are stored to later be forgotten. The *other-than-conscious* selects and categorizes the storage of information based on your beliefs (what *you* consider to be other than average), and those items to which you give special attention.

Knowing this, you can use your selective memory to enhance the process of remembering and recalling information.

Most of us are experts at remembering to forget; this is *selective amnesia*. Without any influence at all, and given enough time, we can promptly forget almost anything. Selective memory is an advantage-- selective amnesia puts you at a disadvantage. The genius uses the natural processes of the mind to set up success. Geniuses pick and choose the information they retain by making their learning experiences "odd and unusual" (memorable) in the mind.

A young mother once prodded her daughter, "Mary," into my office and sat her down in the chair, then turned, looked me square in the eye and said one word, "Help!" Although Mary was a bright and attractive pre-teen, she was regularly failing her tests at school. She did well in the classroom, her homework was always perfect, and when she got home she could review a test she had just failed and get every answer correct. When Mary got into the testing situation, however, her brain would freeze. Mary was creating a selective memory of fear before she even started the test. In her mind she was selectively creating an overwhelming fear of forgetting the answers; she remembered to forget. Her brain was doing precisely what she programmed it to do; "forget" and therefore fail. Although this is not what she wanted on a conscious level, her *other-than-conscious* mind had been well trained to select fear and forgetfulness at test time. My task was to create a new selective memory or mental state for her to access, the memory of "flow." The new selective state had to allow her to receive the answers just as easily during testing as she could at home without the pressure of being graded.

I taught her what my father called the "sponge technique." This was one of the methods that effectively helped me develop my own memory; the one that led to honor-roll status.

I showed Mary a picture of the brain so that she could imagine it like a sponge that absorbs everything she sees, hears and experiences in the classroom. This ability to absorb all information is an absolute truth for the *other-than-conscious* mind. Whenever she needed the information from her brain, which to her was now a sponge full of knowledge, she would say the word, "squeeze" to herself. Instantly the information would flow from her brain into her conscious awareness. Her body was also trained to respond by filling her with the sensations of relaxation until she would feel her hands becoming warm and tingly. By setting up the selective state of squeezing the information from her sponge brain, she became *proactive* toward the test instead of reactive.

Mary was now able to foresee a future of successful test taking, she had rehearsed it over and over in her mind. Mary had always been a very capable student. Now that she had awakened her genius mind with the ability to access the information whenever she needed it, she began to excel. Gradually, over the course of a semester, Mary moved from a C and D student to an A and B student and has sustained that grade level ever since.

SELF-HELP DIALOGUE: RESOURCE GENERATOR

Get into a comfortable position and prepare yourself for access to the other-than-conscious process of "selective memory." With your eyes closed, take a deep breath ...hold ...and then release it. As you do allow the body to begin relaxing. Just let go. You will hear everything that I say as we go through this process. Begin to imagine that every breath is taking you to a perfect balanced state of relaxation. A state that will help you to access positive memories and beneficial skills for the future.

Begin to imagine something you do well--anything at all ... And, as you think of it, begin to get totally into the experience. See through your eyes, hear through your ears and feel what your body is feeling when you are doing something well. And when you are totally there, begin to think of where in your future you could benefit from this positive state of knowingness ...maybe in school ...maybe around friends ...maybe around your family ...But most importantly it will be where you need it the most. Wouldn't it be incredible if this powerful feeling of knowingness could come back with you into your everyday life?...in such a way that negative thoughts and influences would have no control over you at this or any of the awakening levels of consciousness. As you continue to hear my voice, only you will know where the changes will be in the future. So as you begin to take charge and lead a positive, balanced life, you can also begin to think of your future as bright and compelling. Each day a new day with new discoveries for you ...(long pause 2 - 5 minutes) *...Now* *only as slowly as all of those positive changes can be made for you do you return to full awakened consciousness. Slowly now return to the room . . .*

TRUSTING YOUR OUTCOME

In this example the word *trust* is used to help you understand that anything new to you may sometimes feel strange or uncomfortable, but the more you do it the more natural it becomes. If you trust the inner image and the outer actions, then your life can only get better.

Also, trust is used to describe a positive attitude about the new thoughts and actions even if the old ones were to come up. While you're displaying the new behavior, if the old images, sounds or feelings were to move into your mind, you have the perfect opportunity to begin the process of distorting and deleting those old memories so they can no longer influence you.

IF IT DOESN'T WORK, FIX IT!

Who are the people you trust the most? Chances are, you will bring to mind those people who have *earned* your trust in some way: A friend who kept your secret; someone who took the time to help you through a rough period of your life; a person who took extra time out to help you learn something new. Generally, you will trust the people who have proven that you can count on them to be there for you when you need them. These are the friends who are with you through thick and thin.

Who are the people you distrust? Those who have hurt you, taken advantage of you, or otherwise cheated you will probably come to mind. Most people distrust those who they feel have "stabbed them in the back," or otherwise abandoned them at times of need.

Well you can think of your thoughts and behaviors like those people who can be trusted or distrusted. It only makes sense to place your trust in those thoughts and habits that assist you and that will be true to you, and to distrust those that have proven to hurt you in the past.

Of the people on your distrust list, how many are still a part of your life? If you are like most people, there are not many. It is human nature to steer clear of those people who have hurt or harmed you in the past. The same can hold true for your thoughts.

The genius knows how to eliminate, or at least transform, past images, thoughts and feelings that have earned distrust.

Visual learners will find it useful to take any negative or unhappy internal images and place them in the back of their mind in black and white. When the negative images lose their color and energy, they will also lose access to your attention. Soon they will be totally removed from your thoughts because the brain and mind will never perform useless and extraneous tasks. In other words, your brain will not strive to retain or access unneeded and unwanted information. This process will free the visual learner's mind to retain and remember more useful information.

The **auditory learner** will benefit by taking a sound, perhaps a nagging voice from the past, and imagining that it is speeding up, like being placed on a turntable and increased to such a high speed that it distorts and disappears--the mind itself will junk out the useless sound.

The same is true for the **kinesthetic learner**. Even when you are acting as if the new and more positive feelings are becoming a part of you, at times you may reach stumbling blocks. You may need to stop for a moment, acknowledge any negative or angry feelings, and perhaps even turn them up like you would a rheostat (the device in your dining room that will brighten or dim the lights). The neurology of the body will go into overload, blowing out the old circuitry so that new and more desirable circuitry can replace it. In other words, you express the feelings outwardly to make room internally for the new, more appropriate feelings.

A genius learns from mistakes, modifies behaviors and attempts success over and over again until it is achieved. *There is no failure, only feedback.* A genius rejects the concept of failure, becomes aware of feedback and makes the appropriate changes. A wise person once said, "***Those who fail to learn from the past are doomed to relive it.***" This simply means that if we fail to do anything about our past programming it will continue to return to every situation as the dominant belief system out of which all of our behaviors stem.

ACCENTUATE THE POSITIVE!

Let's take a moment to review and learn from some of the behavior change techniques of the past that did not work. We will use two separate examples: 1.) **Old School Psychology**[2] and 2.) **Punishment**. Although still believed to be useful by some, these methods have rarely served as a cure to the problem or even the symptom at hand. I'll explain why.

The brain works on four levels or frequencies: Beta, Alpha, Theta, and Delta. Because the brain is the mechanism that physically stores our memories and influences how we think and behave, it is important to point out how these different brain wave frequencies affect us.

BETA BRAINWAVES

When using many of the old "problem driven" psychotherapy techniques, or when an individual is being punished, we are dealing with the realm of Beta--which is the state of wide-awake conscious awareness. While in Beta, you keep track of your life--you balance your checkbook, pay your bills, decide which errands to run, what you will do for entertainment and so on. Beta is a necessary state of mind; it is where we get things done. However, it is the only frequency that produces fear, frustration, anxiety and self-doubt. Therefore, when therapy occurs at this level of the mind, it may indeed create new programming, but with many of these unwanted emotions involved. In other words, your mind not only takes in and programs information, but also all corresponding feelings and emotions that go along with it. These feelings are what will actually become the initiators of change.

---

[2]By **"Old School Psychology,"** I refer specifically to the outdated problem-oriented methods of "rehashing," the past. I am in no way stating or inferring that modern-day psychology methods are ineffective. In fact, many of today's psychologists will use advanced methods of relaxation when helping patients to deal with stress and overcome problematic behavior.

Let's say you have decided to learn some new information about how to take tests. Would you want to plant that information in your mind with the seeds of fear, frustration and anxiety? Of course not. You would want to sow the seeds of peace, tranquillity and well-being. You would want to sprout feelings of self-esteem, self-confidence and inner worth--all the emotions that will allow you access to every answer on an exam.

So think of Beta as a great starting place, but always remember that behaviors learned through the realm of Beta are stored with emotion. This means that even while you are wide awake and alert, you go about learning and absorbing information even if you try not to. In fact, many psychologists now say that we have absorbed and formed our life experience between the ages of zero and seven. During the first seven years, we are like little sponges soaking in all information about life itself through our parents, family and friends. Then through the rest of our life we are sorting it out--what is true and what is false for us. During those seven years we experience only a small amount of Beta brain wave activity. As children we function in the levels of Alpha and Theta (which you will learn about in a moment), and accept everything without question.

When we program information with fear and frustration, as we do in the realm of Beta, it begins to move slowly back into the *other-than-conscious* mind and into the lower brain wave frequencies, where it eventually becomes what is known as a *Delta Imprint.* In other words, the information, stored with emotion, moves from the realm of Beta into Alpha, then Theta and finally into Delta, which means that it becomes an unconscious behavior. Now the Beta frustration and anxiety has bridged itself into all other brainwave activities and creates the potential for phobias, learning disorders, mental and emotional blocks, frustrations and fears that are almost always unfounded. Once this occurs, we often find ourselves unable to figure out why we are not capable of accomplishing particular tasks or why we respond inappropriately in certain situations.

So the best place to start is by building a base of operation--or a starting point. Instead of going into a therapy session or learning situation and bringing up all the negatives, or the things that need to be improved upon, begin to accentuate the positive. Modern-day therapists are learning to help individuals build up their self-esteem and inner worth so that when they start to face the dragons of their past, they will have the resources and abilities to disable the old beliefs or values that no longer serve them and begin to *enable* the new beliefs and concepts. With an awakened mind, they are able to move inside and place the right sequences and appropriate thoughts into old Delta Imprints, making the results instant and automatic.

When you cut your finger, your body knows exactly how to heal itself. You don't have to think about it at all. Your body knows how to rush the healing white cells to the area. It knows how to form a scab and just when that scab should be shed. The same is true with your mind--it holds a blueprint and knows exactly how to learn.

Just think of how much you have learned since the moment you were born! You had to start by learning how to use all of the different muscles of your body. It's hard to imagine it now, but at one time you had to *learn* how to stand up. It took courage to pull yourself upright, but you finally did it, perhaps with the support of the couch or an end table. Once you were up, your legs may have felt weak, and maybe you weren't even sure that you could stand like the other people around you. But because everyone else could stand, you continued to think and act as if you could too--you grew stronger and steadier.

You were using the most natural form of learning--*modeling*. You imitated your parents, your family members and any friends or strangers who unwittingly crossed your path. In many cases today, children are modeling television characters, which is either good or bad depending upon the selection of programming. (See Chapter Five: The Power of Modeling.)

So you continued to watch and learn. Then, one day something began to happen; you learned to trust in your discoveries and took a step away from the couch. Chances are good that you fell down, perhaps cried, and felt frustrated. But you continued the process until one time you finally pulled yourself up, let go, and felt the support of your own two feet. It felt so good that you stood up again and again. Pretty soon you completely forgot falling down, and only remembered standing. The process of learning to stand is similar to the way in which a Delta Imprint is created. It is a behavior practiced physically and rehearsed mentally until it becomes as natural and automatic as breathing.

Many of the old-school psychotherapy techniques are problem-driven instead of solution-oriented. Now the brain, being a goal striving organism, will attempt to play back the desired changes, but will always get the fears, frustrations, and anxieties which were brought up and discussed during the session. Therefore, the person seeking the therapy gets more of the same, and continues to seek therapy.

As for the person receiving punishment, his brain is being fed conflicting information. He is being told NOT to do something. The brain has no context for the word NOT (remember the red firetruck that you couldn't NOT think about?), so it only has the negative with which to work. Therefore, in most cases, the person will go out and display the unwanted behavior again and again.

ALPHA AND THETA FREQUENCIES

Now let's talk about the next two levels or frequencies of the mind: Alpha and Theta. These are states normally associated with peace, tranquillity, happiness, and contentedness and are usually found in meditation, prayer, hypnosis, in the quiet of nature, or while listening to relaxing music.

As a child, you functioned almost exclusively in the realms of Alpha and Theta. It was natural to you. You hadn't yet learned how to be stressful, anxious, nervous or worried. As you grew older, however, you became a product of your environment--where there is stress everywhere! Now you need to re-train your brain to relax. While your mind eases into Alpha or Theta, you have access to all past information, and from there your mind will continuously create your most probable future based on what your past forecasts. In other words, all your past beliefs and experiences create an expectation for what will most likely occur in the days, weeks and even years to come.

Alpha and Theta can exist only with a relaxed mind and body. Since these realms will then inspire such beneficial feelings as peacefulness and tranquillity, what you would see, hear, or experience while your brain is relaxed will tend to lead into a quick and easy change, usually permanent, within a very short time. Because you created the change while relaxed and peaceful, your brain will anticipate peacefulness each time the new behavior is demonstrated. When you make changes in a comfortable frame of mind, the new behaviors will be comfortable and natural to you. This explains why most successful people take 10-15 minutes each day to just sit back, relax and mentally plan their day.

The Self-Help Dialogues are designed to teach you how to operate in Alpha and Theta. These will work best with your eyes closed so that you can focus your attention inwardly. You will want to be comfortable and relaxed, but remain awake and alert throughout the process.

DELTA FREQUENCIES

The last brain wave pattern is known as Delta, best described as a state of deep sleep and rejuvenation. It is an unconscious dreamland. No one really understands all that occurs in Delta--it is the time when your *other-than-conscious* mind takes over completely. While using the Self-Help Dialogues, you will want to become relaxed and comfortable, but not so cozy that you will easily drift off into Delta.

This brainwave activity, which seems to be unconscious, is actually a learned response. As you progressively activate your genius potential, you will naturally shift from a Beta perspective to the relaxed, natural Alpha/Theta state. It is said that Albert Einstein functioned in the realm of Alpha while wide awake and active. It has been my experience that, with proper training, everyone is able to function in the relaxed brainwave state of Alpha where, like Einstein, they can experience a natural flow of creativity.

It is certainly no happenstance that the description of the Beta brainwave rhythm sounds a lot like the left brain while Alpha/Theta states are described very similarly to the right brain. If you are overly critical in your thinking, you are probably blocking the natural flow of Alpha and Theta. It is also interesting to note that manufactured drugs are available which inhibit Beta brainwave patterns. Wouldn't you agree that it would be much safer and more effective for you to train the brain to create its own natural *endorphins* for relaxation? Endorphins are the actual chemicals released by the brain when you feel good. A natural, healthy release of endorphins is exactly what occurs through the use of the Self-Help Dialogues. Go ahead, enjoy yourself; they are free, safe and non-addictive.

We Always Make the Best Choice!

It is very important to realize that we always make the best choices with the information at hand. This next technique will help you to make changes to the past so you can make more appropriate choices in the future.

Self-Help Dialogue: The Power Of The Past

Prepare yourself for a journey into the past. Close your eyes and become aware of my voice, allow my voice to move between the right and left hemisphere of the brain. For most of you, the right side of your brain controls your creativity ...this side of your brain has no sense of time ...it is the free-flowing side associated with passive characteristics ...feminine qualities ...Imagine with me that this side of your brain is relaxing so you can receive full benefit from it ...and as that part of you continues to relax, become aware of the left hemisphere of your brain. This is the side of your brain that controls the logical functions. This is the analytical mind of you. As you become aware of this part of you, just let it go ...just let it go. There is no reason for you to listen to any other voices ...no reason for you to listen to any other sounds. In fact, from this moment on, all other sounds will cause you to become more aware of my voice and more aware of relaxation as it now moves between the right and left hemisphere of the brain.

It is from here that you can remember a time ...remember a time in your past ...a time that you would like to change ...and as you think of that time see it from a distance ...over there ...and notice what you remember about this past time. As you watch it, over there, begin to imagine that the color is leaving the picture and all of the sound is gone. If there are any feelings left about that picture, just let them go as the picture completely melts away. It is from here that you can begin to imagine a new past, one that will be stored in the same place as the old past which is gone. But, in the new past changes have been made. Some of these changes will be in pictures, some in sounds and yet others in feelings. First, as you begin to imagine a new past you begin to go into the body of anyone else in the experience. It could be your friend or a family member, but just for a moment, imagine seeing through their eyes, hearing through their ears and living life in their body. What would you do differently? Given this unique opportunity, in what way would you respond or talk to the you of the past? Take a moment and experience the memory from all directions. Now notice your reactions of the past. What skill or ability would you need to have in order to handle the situation better? ...to handle it in such a way that you could learn from the experience and benefit in such a way as to become more of what you want to be? When you think of a skill or ability or attitude, take that into your body of the past. See through your eyes with this ability, hear with your ears and sense and feel with your body of the past with this new skill. Begin to breathe the way you breathe when you know something wonderful is happening. Because right now, this moment, you are making a change...it is a positive change ...you are changing your past in such a way as to learn from the experiences and to react differently in the future.

So, as that change in your past occurs, where else in the past could you benefit from this positive behavior? Think of at least 3 times consciously and allow your other-than-conscious mind to think of the thousands of other places and times you could benefit from this positive behavior in your future. Yes ...yes indeed, changes to your past influence your future. The more positive information you place in your mind, the more positive your life becomes. You will find that as early as today those things you do well you will continue to do well ... in fact, you will begin to do them better. The things you feel you need improvement upon, your mind will begin to make all such improvements so that you will soon, and very soon indeed, look back over all the changes, one by one, and notice that the days soon become weeks and weeks, months, and the months, years ...and that all of the new positive changes become natural and normal to you. Some of them happened immediately ...others during the first week ...still others during the years to come. But, one thing is certain ...the changes will be made as you need them to live a richer and more fulfilling life now and in the future.

Now it is time to slowly return to the room ...only as slowly now as all that was said, heard and experienced can become truth for you. Take all the time you need for the changes to your past to become permanent, lasting forever . .

CHAPTER FIVE

THE POWER OF MODELING

Scientists are now discovering that the human mind is far more adaptable than anyone had ever before believed possible. At Stanford University it was proven that even the average student's mind was capable of *advanced modeling.* Also known as "remote learning," advanced modeling is the ability to sit in a room and mentally project yourself to another location and actually learn what is happening there. But I am getting ahead of myself. It would not be appropriate to teach a child to drive a car before she knows how to walk. Yet you will want to be open to all the possibilities as you discover the powerful technique known as modeling.

What do you think learning would be like if you suddenly realized that you already know everything the instructor knows? What a thought! It would probably seem almost as if you were cheating on your homework and exams. Is it possible? Let's find out.

The first university I attended was Ferris State in Michigan. As I headed north with a car load of high school buddies for my first semester of college, I was making big plans for an illustrious college career. By the time we turned onto the campus driveway, I was confident that I was to be their latest and greatest track superstar. I gave little consideration to the education I would receive or the studying required. After all, I could run the 440 faster than anyone in my home town. I was in for quite an awakening as I found out that there was a big difference between accomplishing honor roll status in high school and surviving as even a passing student in college.

I was much more fortunate than most newcomers to college, however. With the training my father had already given me, I knew it would be necessary to improve my memory and self-image if I was to survive academically. I began the process of asking myself the question, *"How am I going to be successful in college?"* I repeated the question to myself several times a day, each time trusting the solution was close at hand. At first, I didn't have a single clue as to what the answer could be. But I knew there was a solution--there had to be, because a problem cannot exist without its companion (the solution). I knew this the same way as I know that the sun rises in the east and sets in the west.

Then one day my track coach told me about a super-learning course being offered at the university. It was a non accredited class and sounded like fun to me, so I attended on my own time. As I look back on it, I now know that the primary skill they were teaching for "superlearning" was modeling. It was not the remote modeling they had discovered at Stanford, but rather a kind of "mental projection" into the mind of the instructor.

I have some shocking news for you -- *Every teacher wants and needs you to succeed.* Remember, your teachers once went through an entire education to learn how to teach you. Most teachers will all but trip over themselves to give away the answers that will be on a test. All you have to do is know *when* they are giving away the answers!

Because teachers are people too, they will have developed patterns and behaviors in their role as "teacher." They are purveyors of information and will tell you all that you need to know. Once you decode their pattern, you will be home free.

First, imagine for a moment that you are your teacher.

1. ***What would you want your class to learn?***
2. ***What would you do to make sure that they are understanding what you have presented?***
3. ***How would you know that your class is learning what you are presenting?***
4. ***What is your intention in teaching this material? What is the outcome you desire for your students?***
5. ***How might your students use the information in their life experience?***
6. ***How might you relate what you are teaching with past information to make it easier to remember?***

Holding these questions in your mind will set up a servo-mechanism, a goal-striving system that works. As you think with the mind of the teacher, you are setting up a mental process for storing and retrieving information. Our minds focus on what we think about most. The genius learns that by dwelling on the solution, all problems can and will be overcome. If you provide your subconscious mind with a goal and give it time, it will figure out the best solution. Modeling is a process of accelerating what we have all learned early in life--the skill of *imitation*.

As small children most of us played "House" or "Work," and in those games of pretend we began to think and respond like our parents, a sit-com family or a friend. The skills of modeling are inherent; they are the foundation of genius. Awakened geniuses are those who have learned to tap that superconscious reservoir of information so as never to waste time or energy in recreating the wheel. They simply make modifications and enhancements until they can move forward with optimum speed and efficiency.

Modeling simply means that you take yourself out of the picture, along with your feelings and emotions, and watch with an unbiased opinion, just allowing it to unfold. It is time to begin trusting in the processes of the *other-than-conscious* mind. For some reason or another human beings began to assume that the mind needs to screen out information. I suppose this would be positive at times, but at other times it is simply inappropriate. This mental screening process is what has made us critical and at times skeptical. When you use your mind to be critical of everything, you are processing all information through the left brain, which uses considerably more energy than the flow of the right brain would. This tends to make you tired and bored with the information.

What a difference it will make when you learn to remain open and receptive like a child. With a curious, child-like mind, you can take the information, apply it to your life experience, and decide whether it is true for you. This in no way implies that you should just accept what others say without question, but rather suspend your judgment and simply allow the information to flow in a relaxed and natural way. Then, if you need the information later, it will easily flow into your mind.

One of the stumbling blocks teachers often encounter comes when they unwittingly begin to assume that the students know the subject matter before they begin to teach them. It's not their fault. They simply know the material so well they forget the student is starting at square one.

Has this ever happened to you? Can you remember a time when you were trying to teach someone how to do something that you knew well? Was it frustrating? Did you get impatient? Were you wondering why they couldn't get it when it came so easily to you?

One of your challenges is to continuously investigate the clarity of the lesson being presented to you. This is why the skill of modeling is so important. If you are ever confused about the information presented, **ask questions**, and ask again, and again if necessary, until you understand. It has often been said that there is no such thing as a dumb question. It is your responsibility to inform the instructor when you are confused. This is true whether you are six or ninety-six.

KEYS FOR MODELING IN THE CLASSROOM SETTING

1. **Sit with an open posture.** *Tests have shown that when your hands or legs are crossed you shut down one side of the brain. With an open posture your brain is able to fully operate.*
2. **Let your mind relax with the information.** *Allow your mind to wonder what it would be like if you could remember everything the instructor was saying. One easy way to monitor your relaxation is to maintain an awareness of your hands. If they are warm, you are relaxed. If they are cold, you are tense.*
3. **Set your mind up with curiosity.** *Ask yourself from time to time, "What is it that I need to learn from this information?" and, "How would this information appear on a test?"*
4. **Take notes.** *Set up your own style of shorthand. Get creative here. Remember key phrases or sections to which the instructor gave special attention, extra time, or verbal emphasis. Example: The instructor emphasizes a particular statement or phrase by continuously underlining or pointing out the information on the chalkboard. (See Chapter Eleven)*
5. **Ask Questions.** *If you are bored with the information, chances are the other participants are finding it dull as well. Ask reinforcing questions when appropriate. As examples: If the instructor emphasizes a certain area of the text, you could ask if this will be an important area to study for the exam. You might ask where you could find other information to expand your knowledge base on the subject. Ask questions that will make the information more memorable, such as, "Where might I apply this information on the job?"*
6. **Take 5 minutes and review information before falling asleep.** *Each night, before you fall asleep, take a few minutes to mentally recall as much of the information as you can bring up. Challenge yourself. With practice you will realize an unlimited faculty for remembering and recalling information. You may amaze yourself! Remember, your brain is like a muscle; it gets stronger with use.*

Your brain has already mastered the art of modeling. I'll give you a perfect example of something almost everyone has experienced. At one time or another we have all watched someone achieve an incredible feat. Perhaps an Olympic champion running the 100 yard dash in record time, or performing perfectly on the parallel bars, or scoring that final winning point. How did *you* feel when that athlete won? Did you feel the exhilaration of a winner? Was your body responding to what you were watching? Maybe your heartbeat quickened, your palms became moist, and your body tensed as you watched those final moments of victory. Did you feel some of that determination and sense of achievement?

Even though you were not a part of the actual physical experience, your body still felt the neurological transfer of all the joy, expectancy and hard work of the other person. If sports isn't your bag, how about this: You are watching one of those good vs. evil type movies and the good guy actor has just reached that triumphant moment when good reigns over evil. Were you there, a part of his victory? Or how about this one: Your best friend has just walked through the doorway and thirty people jump up from everywhere shouting, "Surprise!!" and throwing confetti. Calls of, "Happy Birthday!!" echo all around you. Do you feel the surprise, pleasure and excitement of your best pal? If you're human you do! We all live vicariously through others, and this natural tendency is a vital mechanism of the genius mind. Just imagine what it will be like when you have at your command the feeling of scoring the winning touchdown, hitting a home run, or bringing home a straight "A" report card.

Did you ever stop to think that you can pick and choose your role models? Let's say that Bobby Brains sits next to you in class. You always thought of him as the smartest guy you've ever met, maybe you envied him a little, and he has certain characteristics that you have always admired. Imagine what it would be like if you could tune in to Bobby, the same way that you allow your feelings to naturally meld with the Olympic champ crossing the finish line. Imagine that you could tune in with a full sensory perception (using all five senses) to experience all that Bobby experiences. You would be learning, almost by osmosis, what allows him to be the person that you admire.

Now you will become the curious genius--in a perpetual state of wonder: How do you do what I want to do? What makes you good at what you do? Why does it work for you? How can I learn to do what you have done ...and more? If you were to model just one successful person, and continued the process until you have mastered the way that he or she thinks, acts and responds, your skills would multiply far beyond your wildest dreams.

TAKING A TEST BECOMES FUN

Test anxiety, as it is commonly called, is triggered by a belief that you might not remember what you studied or that you will forget what you know when in the testing situation. There are a couple of causes for test anxiety. First, and most common, is based on the number one rule of the mind: ***The law of mind is the law of belief.*** If a person believes that during a testing situation he or she will not be able to remember the information, then the unconscious mind has been programmed to forget the material. In other words, the person has set up a self-fulfilling prophesy. Marcus Aurelius, who ruled the Roman empire from 161 A.D. until his death in 180 A.D., once said, *"Why be surprised when a fig tree blossoms figs?"* Think of your thoughts as seeds that have been planted into your mind--they will all bear fruit. You can't expect to harvest strawberries when you planted onions!

You can think of modeling as seed planting; it will set up the optimum learning environment. If you keep an open mind and believe that you can learn what your instructor knows, then your *other-than-conscious* will go to work for you building a mechanism that will fill in the details. This means that if you provide a goal to the mind it will search for the easiest completion of that task. Learning and testing will become easy, and when it's easy it can also be fun!

Your mind is a servo-mechanism, which means that it is goal-striving. I think Henry Ford said it best, ***"If you think you can or if you think you can't, you are right!"*** Most people fail before they even begin because they have set their minds on a course of failure. A genius looks at a problem knowing there is a solution. Every adventure movie you have ever seen started with a problem and ended with a solution; for a genius, life is an adventure--life is fun.

SELF-HELP DIALOGUE: MODELING IN THE CLASSROOM

Prepare yourself to be mentally teleported to the classroom of your mind. As you close your eyes, move your awareness to the center of your brain between the right and left hemisphere. Allow the sound of your own voice to become smooth and comfortable to you.

*Take a moment now and think of three things you saw today ...and let yourself go deeper and deeper into relaxation...*pause *...Remember now three things that you heard today and go deeper and deeper into relaxation ...*pause *...Remember three different feelings you have had and just let yourself go deeper and deeper into a positive, relaxed state ...*pause *...Now think of two new things you saw today ...and go deeper and deeper ...Two new things you heard today ...deeper and deeper ...two different feelings you experienced today ....* pause *. . .Take a moment and picture what you look like right this very moment . . .relaxed and comfortable... Take a moment and find out what you are really feeling right this very moment ...and then go deeper and deeper into the relaxed state ...*pause ...

F*rom here I would like you to imagine that you could float on the ceiling and look down at yourself ...what would you look like? ...What positive things would you say to yourself down there? What would it feel like floating up there near the ceiling? ...* pause *...Now take a moment and imagine that you are floating above this building. What would it look like out there today? You know that your physical body is resting peacefully and is perfectly safe while your mind is taking a journey now in the world of your imagination. What would you say to the world out there? ...What would it feel like to be free from all the thoughts and cares of the day? ...free to be anything you want to be ...just let yourself go even higher now ...as if you were looking down from a cloud ...*pause *... feeling yourself going further and further now as if you are sitting on the moon looking back at earth. How beautiful does the earth look from this place? ...Where on the earth do you live? How do you feel about the world from this perspective? Feel yourself turning toward the stars ...and one of those stars is shining a bit brighter than the others. Imagine yourself going there through time and space ...*pause ...

It is here, with the use of your imagination, that I would like your other-than-conscious mind to help you create your own classroom.

What would it be like? ...What would you see, hear and experience? ... pause ...*Take the time in this classroom of your mind to imagine that your teacher is there instructing you. Imagine yourself for a moment in time seeing through your teacher's eyes. What would the classroom look like from this perspective? Do you look interested in the information? What could you do to improve your attitude about learning from this perspective?* ... pause ... *Now take a moment to imagine yourself hearing what your teacher is thinking about you ...What can you improve about the situation so that the instructor holds positive thoughts of you in mind?* ... pause ...

Continuing to relax with the process of modeling ...Imagine to the best of your ability what your teacher is feeling ...Take a moment to mentally review the information that will appear on the test in your mind... That's right ...imagine that your teacher is making a review of the information you will be expected to know on the upcoming test . . . Take a moment to imagine all the probable questions and answers ...It is from here that you will make the mental notes as to where you will need to study the most ...After listening to this Self-Help Dialogue you will find it easy to remain relaxed and motivated to study and learn ...You will find a positive mental attitude ...it will influence you in the morning and throughout your day. It will build the strong positive beliefs that will allow you to learn even in the most difficult situations ...Upon awakening from listening to this Self-Help Dialogue, you will find yourself relaxed, refreshed and revitalized. Take all the time you need and when my voice returns it will not startle you at all. In fact, it will place you in a peaceful, balanced place in consciousness ...and this is so ...(long pause) *And now you can slowly bring yourself back from your imagination and your inner classroom:*

First, back in our solar system ...Then back into our atmosphere ...Then back into our country ...Then back into our state ...Then back into our city ...Then back into this building ...Then back into your body ...And, slowly now you can begin to think how, by modeling, you can make changes in your learning patterns ...Slowly now coming back into the room. Take all the time you need . . .

"The thing always happens
that you really believe in:
and the belief in a thing
makes it happen."

Frank Lloyd Wright

CHAPTER SIX

THE THEATER OF THE MIND

We spoke in earlier chapters about the thoughts and words that are moving through your mind during the course of each day. What do you suppose would be the purpose of this internal dialogue?

Actually, this constant barrage of mental verbiage is to keep you remembering who you are. What if you were to awaken tomorrow morning and you forgot to remember who you were? Would you be able to answer even these simple questions?

What is your name? ...How old are you? ... What do you look like? ... What do you do for a living? ...Who is your family? ...Do you even have a family? ...Who are your friends?...Where do you live? ...Where's the bathroom? !

With this example, you can begin to grasp the significant role that your self-talk plays in your day-to-day life and activities. Yet the above example is only one way in which these internal words are affecting you. Self-talk is also what creates mental pictures and images. Here again, we are getting back to the basic rule of the mind, for it is in how you perceive these pictures that they will either serve you or control you.

Do you want some proof? Here we go:

Pause here and think of a happy time from your past. What's the memory like? How many details can you remember?

If you are like most people, this memory is quite vivid. Most likely, it will come to you with thoughts that are colorful, bright and brilliant in your mind. The sounds of the experience are probably clearly audible to your internal ears and chances are you find it easy to allow yourself to flow with the feelings the memory induces.

How *did* you feel at the time? What made it a joyful experience for you?

How do you respond to the memory now? How does it feel to remember that happy segment of your life? Perhaps your mouth responded naturally by circling upward into a little smile. Did your body respond in any other way? Did you notice your breathing changing, or your heart rate?

Now come back to the book. Since every person is so incredibly individual, this example will vary from person to person. Overall, however, most people probably felt quite comfortable with getting right into the experience and reliving it. You can take a moment to go back over the example if you wish to further understand how your memories and responses are affecting you.

Let's try another example:

Pause here and think of a sad or unhappy time. Make it something that happened a long time ago. What's the memory like? How many details can you remember?

Chances are these pictures are in black and white. Perhaps like a snapshot, not moving at all. There is probably little or no sound in the experience. You are remembering the scene from a distance like watching a movie or looking at old photographs.

You can now come back to the text and notice the differences in the way you store information.

COLOR -- BLACK & WHITE

SOUND -- NO SOUND

IN THE PICTURE -- OUT OF THE PICTURE

MOVEMENT -- NO MOVEMENT

Most people believe that what they have stored in their minds is permanent and unchangeable. They are very much mistaken. To reiterate this point, remember, ***those who refuse to learn from the past are doomed to relive it.***

SELF-HELP DIALOGUE: SHIFTING FILTERS

Prepare yourself for a shift in the way you think about yourself. Close your eyes and become aware of my voice. Notice how quickly you now find yourself relaxing. This time, bring up in your imagination an unhappy time. Put it in a picture frame in your mind and slowly take the color out of it. As you do, notice that it becomes black and white. Any sounds that you are hearing about that past time are getting further and further away as if they are coming from the moon. You know they exist, but they are now coming from outside of your awareness. Any feelings that still remain ...just take a moment and imagine them melting away into the black and white picture. As this happens, imagine that the part of the picture that is white is getting whiter and whiter and begins to fill the whole screen of space. When that occurs, whatever you learned from the experience instantly comes to you ...either consciously or unconsciously.

Right at that moment you can begin to remember a happy time and you place this picture into the frame. As you do, I want you to add more color to the picture. Bring up the sounds of this happy time. Imagine that you are there breathing the way you were breathing then, happy and fulfilled. Imagine looking through the eyes, hearing with the ears, and sensing and feeling with the body of that experience. At just that moment the whole picture in your mind comes to life. It is as if the whole thing is happening now. And, as you go through the experience you can use your mind ...your powerful mind, to help you go back into your past and use this technique of shifting colors, sounds and feelings in such a way that whatever happened in your past that may be stopping you from awakening to your full potential would be changed permanently, changed in such a way as to free up your mind to enjoy the creation of your future now ...right this very moment. For as you take this time to go back and shift memories, placing your awareness on the positive things in life, you will find a new world opening for you ...a new, bright and compelling world. A world that will allow you to express and be who you are

...unlimited. You have started a process of change ...change is the nature of all things. From the spring flowers, warm summers, the falling leaves to the blankets of snow ...change is the one thing you can count on.

Now begin to think about where in the future the changes you are making today will influence you ...some today ...some next week ...some into the months to come. Only you know exactly where all of these positive changes will be. Some you will know about consciously ...some will just happen for you ...through this activation of the other-than-conscious mind ...(pause).

When you feel that you have made the necessary changes in your past and placed the necessary behaviors into your future...and when you are convinced that the changes have been made once and for all ...positively ...you can start your journey back into the room. And, when you return, you can open your eyes and notice the subtle little changes in the room.

IMPACT WORDS

I recently attended a conference on Ayur-vedic medicine, an ancient East Indian method of health care. I was fascinated as the speaker, an American physician, spoke of how the Indian culture integrates the mind/body connection to invoke natural healing. I was indeed entranced as he shared his own personal experiences with patients who had triumphed over cancer and other deadly diseases.

Then another physician, this one an Indian native, took over the podium. He spoke with a soft melodious voice and thick accent. I could barely make out a word he was saying. I was sure that his knowledge and experience were extensive and truly wanted to understand what he had to say. Yet no matter how hard I struggled and strained to understand his words, I could not. Eventually my mind wandered off to other thoughts--I had tuned the speaker out completely.

As I glanced around at the other participants, I saw a room full of polite faces and vacant eyes. What happened? We all very much wanted to learn what this fellow had to say. Indeed, his dialogue may have been absolutely breathtaking, yet he had lost the entire group's attention. Why? Because even though he was speaking English, it was not in a "language" we could relate with.

Have you ever been sitting in a restaurant, minding your own business, and then suddenly found yourself tuning in to the conversation of a couple at the next table? At that moment your ears perked, your body may have shifted toward the speaker, and perhaps you were even straining to hear each word. What drew you toward their conversation?

The people at the other table had probably shifted their discussion to something that held some interest to you and were perhaps using some of your *impact words* as they spoke. You would not have paid a moment's attention to that next booth until they started talking in *your* language.

So let us once again move into the realm of the mind and take a look at these wonderful little communication devices we call WORDS. What exactly are words?

Webster's Says:

WORD (wurd) n. 1.a. A spoken sound or group of sounds that communicates a meaning and **can be represented graphically.** b. A **graphic representation** of a word.

As we notice in this definition, it's a spoken sound or sounds that can be represented graphically. So each time we speak we are representing something that **can be seen graphically**. Wouldn't it also be safe to assume that even our words that we say to ourselves are being represented graphically, but on an internal screen, the screen of the mind?

WORDS AND THE IMAGINATION

Take a moment and close your eyes to imagine each of the words or phrases as you read them, then move on to the text . . .

Imagine a red fire truck ...the door to your home ...your favorite movie star ...freedom ...a pitch fork ...independence ...a police officer

...confidence ...a blue sky ...a golden sun ...honesty ...green grass. You can open your eyes now and bring your mind back to the room.

Were you able to imagine a firetruck, the door to your home, your favorite movie star?

Were you able to imagine freedom, independence, confidence or honesty?

What is the difference in the words?

The main difference is that with the fire truck and your home door you can easily identify with something that is real and tangible to you. The others are not tangible. No two people have the same concept of freedom or think of independence in the same way. There are many of these "intangible" words, and these are one of your most powerful tools in accessing the *other-than-conscious* mind.

There are certain words which house our values. We will call them ***impact words***. These words help run the programs of the *other-than-conscious* mind. How can these words be of benefit to you? First, if you know what these words are, you can use them to create new and more beneficial behaviors with your pre-existing positive programs. Second, you can perhaps find out how these words are influencing unwanted behaviors and then make appropriate changes.

SELF-DISCOVERY: IMPACT WORDS

How do you find your impact words? Take out a piece of paper and fold it in half and then in half again. Use each of the four folded sections to answer the following questions. After you have listed as many as you can possibly think of for each question, fold the paper so that you can no longer see the answers to the previous questions.

1. What has to be present in a job for you to enjoy it?

2. What has to be present in a relationship for you to enjoy it?

3. What has to be present in a hobby for you to enjoy it?

4. What has to be present in your life for you to enjoy it?

Now unfold your paper so you can see all of your answers. Circle the answers that are the same.

These are the words that will have the most impact upon your mind. People are much more responsive when the communication flows within their comfort zone. New ideas and concepts will easily become a part of your thinking when they are brought into your mind in *your language* and with *impact.* Remember, your mind is working perfectly. The genius uses what has already proven to be effective. Geniuses realize that the less they need to do, the better the new patterns will work and the quicker they will become the positive habits they have chosen.

JOE'S PLUMBING

There once was an old fix-it man known to everyone in the town as "Old Joe." Joe could fix anything, but he was especially good with pipes. It was the middle of winter and the boiler in the grade school building would not work. They tried everything, but all of their attempts failed. Finally, in desperation, they called Old Joe. After they told him of all their efforts, Joe scratched his head a few times and walked up and down the length of the boiler. He walked back to his tool box, took out his hammer and walked back down the length of the machine. Then, without any warning, he took his hammer and tapped gently on a valve. Instantly, the boiler started back up and continued to run without hesitation. Joe packed up his tools and went home.

A week went by and the school received the bill from Old Joe. It was for $1,000.00. They were furious to say the least. The matter was referred to the superintendent who telephoned Joe and asked for an itemized statement. How could the cost possibly have been $1,000.00 when all he did was tap his hammer once? Old Joe agreed, and being the simple man that he was, sent a simple statement charging $1.00 for hammer tap and $999.00 for knowing where to tap!

Joe's Plumbing	
Hammer Tap	$ 1.00
Knowing Where to Tap	$ 999.00
Due and Payable	**$1,000.00**

This story is the perfect example of the Master of the Mind. What is most important is to know where to intervene and then to use the right tools. We will soon learn where these powerful impact words can be used and how they will help you awaken the genius mind within you.

SELF-HELP DIALOGUE: IMPACT WORDS

Prepare yourself for a journey to an inner world of communication. Close your eyes and take a deep breath, and as you do hold on to a commitment to your goals.... hold now ...take in a little more air and then just let it go, and with it all thoughts and cares of the day. As each thought, each care goes through your mind just let your higher, creative mind take care of all the details, all the fine points.

*Take a moment and think about your present job (*or *career plans)...How is this job (plan) serving you? How are you serving it? Take a moment now and think of your impact words ...Do they fit into your present job (plan)? If not, what job would give you* (insert impact words) *...As you think of that job in the future, what would you need to do to make it happen? What could you begin to do today to ensure your successful compelling future with all the* (impact words) *you would need? ...What would you be wearing? What would you hear? What would you feel like when all of this has happened? What if, because of all the* (impact words) *your relationships improved? What would it be like in the days, weeks and months ahead? With all these new* (impact words) *what effect will it have on your life today? As you look into the future with new eyes ...seeing clearly all that you will become ...with all this new* (impact words) *what will you be hearing from others that will convince you to continue to use your mind for a positive purpose ...the purpose is to improve your life ...your world ...and your experience. What would it be like if you could just step into a beautiful hot air balloon and cut all ties to the past ...one by one ...and venture out into a new bright, compelling future full of all the* (impact words) *for you that would allow you to soar over the cities and mountains, just as you would imagine a ride in your beautiful balloon. This is your balloon, and it can take you wherever you want to go. So take a deep breath and allow yourself to imagine cutting the ropes that hold you to conscious thinking and begin to feel yourself* *lifting into the clean, fresh air of your creative mind. From this place you can look down at the ropes that held you back ...and you can feel the ultimate freedom of going where you want to go. It is from here that you can continue the journey. Allow your mind to take you on a journey of the inner mind. When my voice returns, it will not startle you at all. In fact, it will cause you to go into a more creative thinking process.*

(Pause)

Now as you return you can imagine landing the balloon outside in the yard, and you can feel yourself walking back into the room where you notice how your body and your conscious mind have been relaxing. You bring all of that wonderful relaxation back with you now into the room as you open your eyes to all the new possibilities ...You can simply glance down and notice all of your powerful impact words.

Taking all the time that you need to return back into the room ...Come back ...Take a few deep breaths ...eyes open, wide awake . .

CHAPTER SEVEN

BUMBLE BEES CAN'T FLY!

When "Jane's" mother telephoned me to schedule an appointment for her daughter, she was exceedingly vague in her description of what Jane needed. All that I could get from her was that Jane was having difficulty performing back flips. I assumed she was a gymnast. When young Jane strolled into my office I became somewhat confused. This was not the sinewy, tiny-framed gymnast I had expected. Rather, she was quite tall and slender; almost statuesque. Long auburn curls fell past her shoulders and framed an expressive face with huge, round eyes.

My first impression was correct, Jane was not in gymnastics at all; she was a cheerleader. My confusion was honest; I had no idea that cheerleading had become such a highly competitive and aggressive sport in high schools and at the college level. Jane knew that she was the best cheerleader on her squad, and she had big plans for her cheerleading career. There was only one thing that stood in her way -- the ability to perform a consistent back handspring.

When I started interviewing Jane, I asked her to go inside and imagine what it was like when she knew that it was almost time to perform a back handspring. Immediately her face turned to ice; an expression reflecting the sheer terror in her mind. I asked her to stop the picture in her mind and tell me what she was seeing. Her thoughts had taken her to an incident that occurred years earlier when a back handspring ended in an incorrect landing. The pain had been intense and the injury laid Jane up for several weeks.

I now knew that Jane was thinking from the past perspective. Even though it is a different time and gym, and she is now a much stronger person, she still approaches each new back handspring with the memory of the past. It had become a loop in her mind; something like a VCR set on continuous replay. Each time she attempted a back handspring, her mind would loop the images of failure after failure.

I started Jane's session by helping her to delete and distort the old image. Her *other-than-conscious* mind was then ready to accept that it was no longer necessary or relevant for her to continually re-experience that past trauma. I asked her if she had ever been able to do a back handspring correctly. Her face beamed. "Of course," she said, "hundreds of times."

It was one incident out of hundreds standing in her way. She had created the self-fulfilling prophecy that she would fail over and over again based upon her past thinking. I helped Jane to understand that if she continued to dwell on that past, she would continue to get more of the same situation. Jane assured me that she was ready to let go of her past experience. I helped her to continuously distort the image and belief until she was no longer able to recall the incident of failure.

Jane still knew that she had failed at one time, just as we all have failed many times while learning to walk, run or even ride a bike. There is always a series of failures that go along with the successes of life. Now Jane's mind was in a perfect place to begin focusing on what she did well. I asked her to think of a gymnastic process that she could do rather easily. She rolled her eyes. "I can do a cartwheel forwards, backwards and with my eyes closed," she said.

"What does it feel like when you're doing a cartwheel?" I asked.

"Well," she said, "I don't even have to think about how to do a cartwheel, I just do it. I can do ten in a row, no sweat. I just think of where I want to go, and I go."

"What would it be like," I asked, "if you could begin to think that same way about the back handspring?"

Jane grinned. "That would be wonderful!"

"How many times do you think you would need to do that back handspring before you could become convinced that it's possible for you today?"

"Well," she said thoughtfully, "if I could do it just once, I think that I could do it again." Jane hesitated a moment. "But," she said, "I would have to feel confident about doing it."

"That's exactly right, Jane," I said. "If your body can do anything once, it can do it again."

We now worked together to build a new and more powerful Jane--the Jane of the future--fully capable of what she wanted, consistent back handsprings, instead of what she didn't want, the failure, injury and falling down that her mind had dwelled on in the past. Jane created the positive picture of the future in bright, living color. I even had her place her favorite music there. As Jane settled into the process she began to sense the confidence growing within her. At the end of her session, Jane spoke up with enthusiasm. "I think I will be able to do this right away, today!" She announced.

I asked Jane to wait and take some time to rehearse the back handspring in her mind. I also instructed her to replay mentally the perfect back handspring five or six times every morning before getting out of bed and to see herself performing it perfectly in several different places. "When will be the earliest time you will be able to practice the back handspring?" I asked.

"I could do it next week at cheerleading practice," she said. Then her eyes lit up. "Or, I could do it at home, on the front lawn."

Jane left my office with a broad smile. Before starting my next session, I made a note on my calendar to place a call to Jane in a week. I was curious as to just what her progress would be. Later I learned that the phone call would not be necessary. My office staff informed me that after her session, Jane had been demonstrating back handsprings for them right out in our front lobby. Once Jane acknowledged and then released the past image and realized the future possibilities, her future far outweighed her past and became the predominant thought. She simply learned to *do* her action, instead of *thinking* her past.

Most people who are experiencing problems in their life or who need or want to make personal changes find that the same old obstacles appear again and again. There are two fundamental reasons for this. First, they are in the habit of blaming others for their misfortune and, secondly, they simply have no idea of how to make the changes or even where to begin.

Our brains work on a very firm principle. All that we see, hear and experience is recorded as fact and in some way influences us. That influence depends greatly on the amount of attention given the information, either consciously or unconsciously.

Let's take it one step further. If you want to make a change in your personal life but are blaming your problem on another person, your mind will seek ways to prove that the other individual is at fault and will do nothing to change your own personal behavior. Therefore, you will continue to get the same outcome. On the other hand, if you take full responsibility *(even if you consciously still cannot believe you are at fault)*, your mind, which responds to *all* related information, will begin to adjust your personal behavior. As you begin responding to other people differently, their response to you will also change. Soon, as you make adjustments and changes in your perspective, the old habits and patterns dissipate. This process will be simple to you, once you know how, and it is one of the most powerful tools available for any time you may want to change or improve your life.

All of us know people who are constantly complaining about life and the way the world treats them. We also know people who are happy-go-lucky and things seem to go their way. Could it be that these "lucky" people are using their internal genius to make the best out of any situation? If this is true, then those people who always seem to be in the right place at the right time might simply be recognizing opportunity. The complainers are unable to see the options because they are using their full mind power to perpetuate a negative state. Some people are so trapped into negative thinking that they would walk right by a money tree because they are looking down at the ground chanting, "Poor me ...poor me ...**poor me!**"

PERCEIVING LIFE

Take a moment to look down at the picture below. As you glanced at the picture, did you think of the glass as being half empty, or half full?

This exercise may seem somewhat a cliché, but it will help you to understand what your first perceptions have been, and the way in which you are either filtering out life or allowing it to flow into your awareness. Are you thinking with a limited thought process--that there is not enough--something like Jane's belief in the pain and failure of the past? Or, are you thinking that there is plenty of everything to go around--the way Jane felt after awakening her mind to her true abilities and self-confidence.

This in no way means that you shouldn't think of all the possibilities and outcomes. It is to look at the situation more fully and find out where your thinking is taking you. Every successful business planner will work out the worst case and best case scenarios and then plan along the middle road. The successful business has *all* the information so they know what to dwell upon and what to avoid.

According to all of the laws of aerodynamics, bumble bees are incapable of flight. They are much too heavy for their wings. Similarly, science has proven that human beings are not supposed to be able to stand on two feet and walk as we do. But every morning a miracle occurs -- you get out of bed, plant your feet firmly on the floor, and walk across the room. It doesn't matter how groggy or tired you may feel, your feet will carry you wherever you want to go without so much as a conscious thought. Bumble bees take flying for granted; most people wake up every morning assuming that their legs and feet will take them wherever they want to go; "Jane" now approaches each new back handspring with the presumption that it will be her finest; ***geniuses prepare for the worst and assume the best.***

LEAD, FOLLOW OR GET OUT OF THE WAY

This next exercise is designed to help you find out where you are and where your thinking has been leading you. It will also help you to imagine more fully where you want to be and what practical steps you can take today to bring what you desire into reality.

Take out a piece of paper and fold it in half. On one half of the paper write the heading, "The Me of the Past," and on the other side write, "The Me of the Future." Then fold the paper so that you can see only one side at a time.

THE YOU OF THE PAST

Focus your mind on an attitude or behavior from your past that you would like to change.[3] Write it at the top of the page.

Use your mind and imagine the you of the past who displayed that behavior then write all that you can remember about it.

1. ***What did you look like at that time?***
2. ***What age were you?***
3. ***What did you talk like?***
4. ***What did you feel like?***

Remember just what it was like to be that you in the past acting out this particular attitude or behavior.

---

[3]Have your mind catalog this attitude or behavior for use in the upcoming Self-Help Dialogue.

Allow your mind to now drift into the future with this behavior of the past. Write out all that you can imagine your life would be.

Where is the thinking of the past taking you?

Where will you be a month from today if you continue with the old thinking patterns? . . .

Six months? ...

One year? ...

Five years?

Chances are life will be much the same as it is today. Remember the principles of the you of the past --

If you continue to think what you have always thought, you will continue to get what you have always got!

-- and perhaps even worse. Your body may even begin to react to the day-to-day resistance and frustration and create what is known as *psycho-somatic* illness. These are sicknesses and disorders felt in the body, but created by the mind.

Remember that your mind works with rehearsal patterns, which are influenced by *positive intentions.* Deep down at the core of this old thinking something positive is happening ...not at a conscious level, but rather somewhere deep within the *other-than-conscious.* It is this underlying positive intention that we want to use in the next segment of this exercise. It is designed to help you create new behaviors, attitudes and beliefs that will build upon your perfect memory, concentration and your ability to remain focused, even in the most difficult situations. Geniuses know how to use their mind to get the results they want.

The genius has trained his mind to learn along the lines of personal *responsibility.* This simply means that you *respond* to your *abilities*. You are responding with the understanding that you may have created your past actions and beliefs, but you are no longer bound by them. You are stepping out of the prison of your conscious thought patterns and stepping into the free world of the genius -- the world of your imagination. This is where you will build a bright and exciting future. As you begin the exercise, be prepared to activate that inner genius and allow it to become reality for you.

RESPONSIBILITY = *RESPONDING* TO YOUR *ABILITIES*

Now flip your paper over so that you no can no longer see the You of the Past and can focus on the You of the Future:

THE YOU OF THE FUTURE

Take a moment to think of what you would want to be like in the future -- a no limit person, a true genius, fully equipped to use your inner screen of space to create anything needed or desired. You have awakened the power of your *other-than-conscious* mind. If you place a goal within its awareness, it will bring about the results. As you read through the following questions, allow your mind to place a full moving picture into action; then write out all that you are seeing, feeling and experiencing.

What skills would you need to be the kind of genius you are imagining?

What abilities would you want?

What would you look like?

What would you say to yourself if you were there, in the future, looking back over time?

What would your breathing be like?

How would you know that you are there?

How would you know that this is your bright and exciting future?

What would convince you to live this life?

What would your surroundings be like?

Allow yourself plenty of time here so you can get into the experience fully. As you think over what you have just experienced, allow yourself to notice what the first step will be in bringing this future into reality.

So, how will you know that you've arrived? How will you recognize that morning when you awaken, your feet hit the floor, and you are convinced that you are equipped and ready for attaining your goals?

Now unfold your paper and review both sides. As you examine both possibilities, begin to notice the differences between them. Does the past you now seem just a little bit dimmer? How about the future you, is it a bit brighter? Think about the way you talk to yourself about each and notice the differences. Begin, in your own way, to notice your feelings about both the past and future you.

How do you bring it together to release the past and accept the new you? This is where your *other-than-conscious* mind comes into play. Begin to think of the past you with the mind of the future. As your future mind makes the changes, you can begin to notice that the past you is melting away and all that remains is the you of the moment with all the possibilities of the future.

As a genius you can now think of your past as a training ground. The decisions and choices of the past will no longer affect you, as long as you remember that you have always done the best you could with the information at hand. As an awakened genius, you are now learning to place your attention on what you want. You will soon discover that whatever your positive intentions of the past were, they will now be fulfilled in a way that is more appropriate for the *you of the present.* From this point of view you can realize the truth in the statement, *"What You Place Your Attention on Grows"*

SELF-HELP DIALOGUE: BRINGING IT ALL TOGETHER

Prepare yourself to know the you of the future today. Close your eyes ...and in one hand place the ***you of the past*** *...all of the fears, doubts, resentments and faults of the past ...be honest with yourself. Now get into the experience fully ...be there in your past. When you are there, slowly close your fist down ...that's right ...*(Pause)

Slowly now, open up your hand and become aware of your body ...The breathing of your body...the relaxation of your body ...Now, in your ***other*** *hand place the* ***you of the future*** *...make this a more positive you ...use your mind to place into this you of the future all skills and abilities that you would need, want or desire ...begin to breathe the way you would breathe ...as if all of this has happened. When you have it all mentally, step into the picture ...use your imagination to see through the eyes of the future, hear through the ears and sense and feel with the body of the future. When you have done all of this, slowly close the hand which represents the you of the future. You're doing perfectly ...that's right. When you have done that, let your hand go loose and limp.*

And once again, become aware of your body ...the state of relaxation you are now experiencing ...your breathing ...rhythmic, natural and relaxed. Now slowly begin to clench both of your hands and think of the past you and the future you ...and when both images are in your mind, let the hands go and begin to use your mind ...your other-than-conscious mind ...to integrate all the positive intentions of the past with all of the new behaviors of the future. And as this occurs for you ...in your own unique way ...let your mind drift back and forth from the past to the future, accessing all the information it will need to make the changes you desire ... Some of the changes will be conscious ...you will simply do new things. Some of the changes will be unconscious ...just when you need them to receive the most benefit from these discoveries.

As your other-than-conscious mind continues to merge and melt with the you of the past, you can begin to notice the subtle shifts in your awareness. All things in the universe are subject to change. This is the one and only thing we can count on ...CHANGE ...So, as you use your mind to focus on the changes you desire, you can become aware of your hands. Perhaps you have already noticed that your hands are touching. You can allow your hands to slowly move together ...but only as slowly as all the positive changes can be integrated and accepted by your conscious mind. Take all the time you need. But when the hands come together, either now or in the future, you will begin a program of building a bright, compelling future ...the future you desire ...and this is so . . .

(Pause 60 seconds)

And as you now come back into this moment, you will come back only as slowly as all that you have seen, heard or experienced can become truth for you ...take all the time you need ...the seconds are like hours, the hours are like days and the days are like weeks ...slowly now ...returning to the room with all the new skills and abilities working for you ...

(60 second pause)

And as I now count from one to five, all the time you need will have occurred . .

1. *hearing the sound of my voice . . .*
2. *coming back fully into the room . . .*
3. *feeling natural and normal in every way . . .*
4. *with a perfect memory and recall of all that was said, heard and experienced ...and*
5. ***eyes opened,*** *wide awake, feeling better than ever before. Ready to take that first step into your bright, compelling future ...and this is so* ...**and so it is . . .**

CHAPTER EIGHT

YOUR MAGICAL BODY

Up to this point we have discussed your mind and your emotions, but we have talked very little about your physical body. What part, if any, does your physical body play in all this genius stuff? It plays a very important role. Your physical body is the vehicle of expression for your mind and emotions; it is what allows you to function on earth. But let's not mistake the body as the total you. You are more than your mind, you are more than your emotions, and, of course, you are more than your physical body.

Have you ever stepped outside on a starlit night? It's a beautiful sight, wouldn't you agree? The vastness of space, the endless possibilities; at times it can be almost mind-boggling. Let's take a short journey to another place just as inconceivable--the human body. Imagine for a moment that you could look inside of your body with a microscope. Take the power of the microscope to ten times smaller than you are now. Imagine how you would view this magical mechanism you call your body. Think how incredible it would be to view your heart as it pumps blood through the different arteries and veins, feeding the cells and organs. It all happens in perfect harmony. Imagine the respect each cell has for the functions of the others. Take a moment to notice the relationship between all cells, systems and organs and how they move together like a finely tuned orchestra. Notice the synergistic relationship they have, one to another.

Imagine you could now take the microscope another ten times smaller. Perhaps the organs begin to lose their shape, but the cells start showing their neurons and protons. Notice how the crisp, clear distinction that was there before is now gone. All that remains is a world of whirling electrical particles spinning in and out of view at faster-than-light speed.

Imagine you could go ten times smaller again and you are truly seeing into the vastness of your body -- like a scientist looking under an electron microscope. It is said that at the atomic level of our bodies there is a greater expanse than in interstellar space. This means that you are more nothing than something! What is that "no-thing"? As you awaken your genius you will come to understand this *other-than-conscious* power that not only keeps your body operating in harmony, but keeps the sun and all the planets in our solar system and all the other 600 billion galaxies in perfect harmony. You will discover, as those great geniuses of the past have, that we are far greater than we have been led to believe and that a superconscious power greater than our limited conscious mind is in control.

PHANTOM LEAF EFFECT

In the late 1940's Soviet electrician Semyon Kirlian perfected a type of electro-photography capable of producing images of energy fields. With the use of Kirlian Photography, scientists have shown that when a section of a leaf is cut off, the electro-photograph still reflects the image of the part of leaf that had been removed. Even the minute details of the leaf appear intact on the photograph. This photographic "effect," has come to be known as the "Phantom Leaf Effect."

Doctors have also noted a similar effect on humans with a missing limb. There have been cases where an arm or leg is lost, but the person still experiences discomfort as if that limb were still attached. This is called the "Phantom Limb Effect."

This "effect" helps to confirm the statement that we are more than just our physical bodies ...much more. A part of us is keeping harmony and balance within our bodies. It is this superconscious (*other-than-conscious*) part that is continuously at work taking care of all the details. Think of it this way: What if you began to tap into the seemingly infinite intelligence that controls the beating of your heart, your breathing, and the monitoring of all unconscious bodily functions--that which seems to hold this and every world together? What if the possibility existed that everything that "it" knows, you could know? Well, that is exactly what a genius does. A genius is awakened to the possibility.

In the movie **Star Wars,** Luke Skywalker used what the Jedi called "The Force" in his adventure of good over evil. It was their term for that seemingly unending flow of intelligence, and it was what gave him the ability to triumph over Darth Vader, even against the worst odds. This source, this power for good, doesn't care what you call it; it simply wants to be used. But you must train your mind to access this connection in a relaxed state.

With this common thread of intelligence, you will build a link to a greater perceptual reality. A reality where at a meta-level (beyond a level of conscious thinking) we are all connected--where we are all tapping into the same super-computer, and it downloads the information you need as you need it the most. The only thing stopping this flow is your conscious fears and doubts. That is the primary goal of this text--to awaken within all people the understanding that they already know all they need to know as they need to know it. This is the primary lesson in all learning: it is not to "learn" the information, but to recognize what you already know.

THE BODY IS A VAST CHEMICAL LAB

That's right, somehow your body is maintaining a perfect balance between all its major systems. From the digestion of incoming food to its assimilation, to the building of new cells, an incredible number of chemical transactions occur. A genius knows that the food they consume will soon become the body they will wear.

Believe it or not, some people out there think they can eat garbage foods and still have a healthy body. They may think they have some cosmic secret of mind-over-matter, but it simply isn't true. If you build a house with less than superior woods and metals, the house will be prone to early decay and will lose its value quickly. The same is true with these beautiful and efficient bodies given to us for life long use. If you take care of your body, your body will take care of you.

A genius knows the benefit of exercise and creates the time for exercise. It is said that after the age of 27, the body starts to lose a pound of muscle to fat each year. A pound is a pound, but a pound of fat takes up 5 times more space than muscle. The law of nature is change. As of this writing, no one has found a way to stop or reverse the aging process, but research has proven that exercise is one of the best ways to keep your body in tip-top shape. A genius also recognizes that a body free of fat and toxins is an enjoyable body to live in--one that has plenty of energy for any activity in which he or she may choose to participate.

THE BODY ELECTRIC

Your body is also a very powerful electric station and conductor of electricity. This is why some people use the expression, "I'm full of energy," or "I'm tired, I've lost my energy." Geniuses tap into this unending flow of energy and use it to accomplish their goals.

This is very important because the electric current that is a part of you moves through your body and on some days may seem stronger than others. The reality is that the energy never diminishes. It is self-perpetuating when the body is in balance and in harmony with its environment.

A genius knows the benefits of love and the ability to accept and give love. It has been said that love is the combination of all emotions. It has also been said that love is all that there truly is. Geniuses realize that they must have love for themselves as well as love for others. If love is too strong a word for you at this time, think of unconditional acceptance. It is a more left-brain way of stating the same thing.

Geniuses recognize when they are outside of the state of love. Their energy diminishes and they start to think with the belief of lack or the consciousness of "not enough." The best way to balance this effect is to stop yourself, focus your attention on the forehead, take a few deep breaths, and say the words, "I am in the flow." Something wonderful will happen as you use the techniques outlined in this text. Doing so will awaken within you the higher faculties of thought ... *Love is all there is. There is just enough of everything for everyone. You will have all the information you need as you need it the most.* Soon you will find that these and other positive thoughts will simply flow into your mind without warning. Then you will know that you are stepping onto the path of the genius.

This brings us to the nutritional aspect of awakening your genius. We are not going to get into what specific foods you should or should not eat. Rather, we are going to discuss two categories: 1) Fresh and Alive; and 2) Dead and Lifeless.

FRESH AND ALIVE

Fresh and alive foods such as fruits, vegetables, whole grains and clean, clear water bring balance and harmony to our bodies. Because the body is bio-electrical, these "alive" foods are what the body needs. This simply means that the biological system interacts with the electrical system to produce you and all that you are able to express here.

DEAD AND LIFELESS

Dead and lifeless foods, such as refined sugar, white flour and those highly processed and packaged, have a negative impact on this bio-electric processing because they load the body with toxins and lack the nutrients our bodies crave. Dead and lifeless foods are energy zappers.

Your incredible body can take food and water in, break it down to its smallest atoms, use what it needs and eliminate the rest. When studying or learning anything in life, it is important to remember that we process life through our senses and it takes mental energy, which means the foods we consume are broken down and used as energy. If you eat dead and lifeless foods, you are not going to be as swift and precise with your thinking as you will be if you choose fresh and alive foods.

When I was younger I was blessed to have a mother who took the time to investigate the link between diet, health and intelligence. She suspected that unhealthy foods could have been the major reason why our family seemed to be experiencing a lesser intelligence, and she set out to prove it.

When my brother Michael, the eldest of nine children, started to school, each of his successive teachers quickly grew very frustrated by his seemingly unending exuberance and his brief attention span. One innovative teacher would actually have him run around in circles in the classroom until he was tired out. Then she could start the class. The principal of the school as well his teachers wanted what was best for him, but they had to do what was best for the rest of the class as well. They encouraged my mother to put Michael on medication to slow down his hyperactivity. She emphatically rejected this idea as she believed that there must be a healthier option. It didn't take much research for her to conclude that sugar was the most likely culprit. She set herself about the task of clearing our kitchen of all refined and synthetic sugars--not an easy task for a household of the 60's, when adding sugar to everything was the norm.

So began our family quest for the healthiest diet. The children in my family, including myself, went from the special classes for kids with learning disorders to students at the top of the class with honor roll achievements year after year. I consider myself fortunate, for at an early age I was living proof that a strong and effective mind starts with a healthy and natural diet. As a student and an athlete, this was a lesson I would never forget.

GENIUSES KNOW THAT THEY ARE "BIO-ELECTRICAL"

Fresh and alive foods are essential to good health because you are not only a biological specimen, you are an electrical organism. Your bio-electric system has this incredible ability to transform the food you eat into energy. The less effort your body expends in converting food into energy, the more energy is left for you to use in getting things done. This is a key ingredient for anyone wanting to awaken their genius. If the body and mind can spend less time in healing the damage from improper consumption or in attempting to upgrade low-level foods, there is more mental energy left for creativity.

At this point, it is important to stress vitamin and mineral requirements. We live in a society that consumes dead foods--processed, chemicalized, over-cooked and devitalized. "Quick" and "Instant" foods are common. Prepared foods are now the norm for our children. These foods are lifeless and devitalized. They lack energy and eating them puts a great deal of stress on the body. Even foods that were considered healthy in the past are now covered with pesticides and herbicides.

Therefore vitamin and mineral supplements are very important. A multi-vitamin that is natural and chelated is recommended by most physicians. Chelated means that it is prepared for the best absorption into your blood-stream. Natural, chelated vitamins will allow the body to take what it needs and eliminate the rest. More is not always better. When it comes to vitamins and minerals, remember that the body is a chemical factory and it sometimes will store what is not used. Your body needs the proper combinations at all times because it is always building.

What are you doing right now, this very second? Most people will say, "Well, I'm reading a book." True, but in the same moment you read the first sentence of this paragraph, you also built more than 50 million cells. Yet you were completely unaware that it was happening. OOPS, there you go again, building 50 million more cells--each second! This is your body's method for continuously replacing the cells of your body. Recognize this as the true genius that exists inside of everyone. What a miraculous part of you! Pure genius! What else could keep your body alive and control the functions of each system, organ, and cell, right down to the individual molecules and atoms that make up who you are. How else could your body perform such dramatic feats as healing a cut or overcoming a virus? Imagine the intelligence capable of making the kinds of upgrades to the human form that archaeological digs now show have been occurring through eons of evolution. Just think, all this intelligence is right there, within you!

A Genius Avoids Dead Foods -
- THEY LOWER ENERGY

Now let's talk for a moment about dead foods: sugars, starches, white breads, white rice, high-fat meats, processed foods, chemical additives, preservatives, artificial flavorings, caffeine drinks and soda pop all fit into the category of dead foods.

Have you ever heard of the "sugar blues?" This effect from refined sugar is very real. It has been scientifically proven that 20 minutes after taking in sugar: candy bars, cookies, or sweets of any kind, your blood sugar level drops lower than before. This drop creates the craving for more sweets. Have you ever felt that once you started eating sugary, sweet foods, you just couldn't stop? The sugar blues is the culprit. Also, because the brain lives solely on blood glucose (sugar), when an individual has low blood sugar, the brain is not getting all the energy it needs.

Yes, it is true that our body needs sugar, but not in the form to which we are accustomed, i.e., cakes, pies and sweet rolls. Our bodies know just how to convert the healthy and natural foods into all of the sugar (glucose) that is needed for health and energy. It is indeed true that our body converts the food we eat to sugar. So how do you get the sugar your body needs without eating sugar?

Fruits are an excellent choice and the best way to get the natural sugar we need. Fruits are also a natural cleanser for the digestive system and should not be eaten with any other food type.

The best time to consume fruits is in the morning before anything else is eaten or two to four hours after your last meal.

Brain Foods

"COMPLEX CARBOHYDRATES"

Whole Grains - *Rice -Wheat - Millet - Barley - Oats*

Grains are also a sugar, but they burn slowly in the system and offer the most benefit when eaten in combination with vegetables. Grains are also considered "brain food." Most people don't realize it, but their brain uses up most of the energy of the body. If you have a mentally taxing job, you could become just as fatigued as someone with a job involving physical labor.

These grains balance the system and are high in fiber. They absorb some of the sugar and are used later in the intestinal tract for the process of elimination.

UNUSED CARBO'S

Avoid
"Enriched/Refined Carbohydrates"
White Flour - White Sugar

Carbohydrates are also a form of sugar in the system, and it is important to understand their function. Carbohydrates can be converted rather quickly into energy. Therefore if your job involves physical labor, you can eat carbohydrates and not gain weight. This is because they are being burned up quickly. However, your body will store carbohydrates if they are not used. Yes, you guessed it . . .they are stored as fat. The secret is to find your unique balance.

The refined and enriched carbohydrates such as white flour, white rice, macaroni, and other processed carbohydrates are empty calories. In other words, they supply calories but little or no nutrition or fiber. Your body must work double time to reap any benefit from these foods.

PROTEIN AND THE BEEF CONSPIRACY

It's at this point that people will frequently ask, "What about protein?" The concern over protein has been a controversial issue for the past several years. What part does protein play in a balanced program? Well, protein is important; in fact you need protein in your diet to stay alive, but not nearly as much as the Beef Council, the American Egg Board, or the National Dairy Council would like you to believe.

A FEW SHOCKING FACTS ABOUT MEAT

Most people are unaware that it takes from six months to a year for red meat to fully digest. Heavy meat eaters will have undigested meat in their bowels at all times, sometimes as much as four to six pounds. What's more important to remember is that the body is not designed to assimilate meat. If it were, we would have much smaller intestinal tracts, like the carnivorous animals. People who eat red meat may eventually gain some benefit from it, but it puts enormous stress on the body.

When a man or woman has a heart attack, the first change the cardiologist will make is to place the patient on a vegetarian diet. Why not start now, before you become the heart patient? Consider this: A Yale University study of endurance was conducted by a Professor Fisher. He asked a group of vegetarian office workers and a group of meat-eating athletes to hold their arms extended for as long as they possibly could. The results were startling. The vegetarian office workers held their arm outstretched for an average 64 minutes. The meat-eating Yale athletes? Only about ten minutes. This makes the vegetarians, who were not trained athletes, over six times stronger in the area of endurance. Dr. Fisher's conclusion, "Meat-eating and high protein diet, instead of increasing one's endurance, has been shown like alcohol to actually reduce it."[4]

I grew up in "Cereal City" -- Battle Creek, Michigan. This is the home of the Kellogg Company, world's largest manufacturer of breakfast cereal. People who are not from Battle Creek know very little about how cereal was "invented." Dr. John Harvey Kellogg was a physician and researcher. Early in his career he noticed the relationship between diet and disease. His studies brought him to the conclusion that a vegetarian diet, one that was high in fiber and low in fat, seemed to help his patients recover not only more rapidly, but more completely. Dr. Kellogg ordered a high-fiber grain cereal for every patient in his hospital. The rest is history.

Dr. Kellogg quotes the following from the International Scientific Food Commission: "No absolute physiological need exists for meat, since the protein of meat can be replaced by other proteins . . ." It was Dr. Kellogg's firm belief that a healthy high-fiber, low-fat vegetarian diet would ward off disease. It's sad that the successors to Dr. Kellogg lost sight of his original vision when they discovered the profit margin in sweetened, processed cereals.

Also misleading is what advertisers tell you about milk. They would like you to believe that humans can get benefit from the cow's milk that is sold at your local grocer. Think about it, we are the only species that continues drinking milk beyond infanthood and the only species on the planet that consumes the milk of another species. If you were to give your neighborhood cat a drink of the milk you have in your frig, it would probably have a nasty case of diarrhea the next day. Did you know that if they fed the homogenized milk that we serve our children to a calf, it would die in a few short weeks?

What does this prove? That our bodies are incredible! They can even take sub-par materials and dead foods, and still produce results. But, these are not always the results we want. The total genius, one who is prepared for life in the twenty-first century, plans to have a healthy, strong body. And if you are honest with yourself, you already know that your body is a combination of the food you consume, the liquids you drink, as well as the attitude you hold. If you continue to neglect your body by feeding it dead and devitalized

---

[4]Iriving Fisher and Eugene Lyman Fisk, **How to Live**, Funk & Wagnalls, p. 251.

foods, you can be assured that it will discontinue giving you the results you want.

So what about protein? How are we to get the protein we need without eating meat? It is true that you need protein because it will build and repair tissue. However, there are many natural sources from which you can obtain your daily supply of protein. Unless you are doing some form of strenuous physical activity, you only need about 50 grams per day. This is about ten percent of your total calories. You only need ten percent fat intake as well, which can be readily consumed in grains, vegetables, legumes and even fruit. The other 80 percent of your daily diet should consist of the complex carbohydrates. The best form of protein for the body is through vegetables. One potato a day would give you all the protein you would need for that day.

If you have some extra pounds that you want to lose, I recommend that you start today. Remember, what the mind dwells upon, it must become. If you focus on the weight, you will get more of the same. Forget about dieting and skip the calorie counting methods that continuously remind you of your weight and keep you focused on food. These methods have somehow lasted for decades, but they simply do not work. It doesn't matter how many calories you consume specifically, but rather what type of calories are consumed and how your body can assimilate them. In other words, focus on quality, not quantity.

Again, the best rule of thumb is this --

IF IT'S FRESH AND ALIVE IT WILL PROMOTE HEALTH

IF IT'S DEAD AND LIFELESS IT WILL PRODUCE MORE OF THE SAME

When you focus on eating for health, your body weight will naturally return to what is normal for you. It is always safe to say, "Let your conscience be your guide." If we are honest with ourselves, we all know what is healthy and what is not.

THERE ARE SIX LAWS OF NUTRITION:

1. **First and foremost, under-eat.** It takes up to 15 minutes for the stomach to send a signal to the brain to tell you that you're full. In the case of food, less is more.
2. **Never eat and drink at the same time.** Your body is set up to digest solids and fluids separately.
3. **Eat fruits separately so they can work as a cleanser.** They will frequently bind you up if eaten in combination with other foods.
4. **Vegetables are builders.** They are the fabric that forms healthy bones and tissues. Raw or under-cooked is best.
5. **If it comes from a can, bottle or wrapper, it is probably not for human consumption.**
6. **Water is the elixir of life.** The more, the better.

If you remember these six easy steps, your thoughts will be sharper, your body healthier, and your energy will truly be unlimited.

EXERCISE - - IS IT REALLY NEEDED

TO AWAKEN YOUR GENIUS?

You don't have to be a rocket scientist to figure out that a body polluted by toxin-forming foods will affect the mind's functions. It would be logical then that a toxin-free body will feel healthy, have more energy and a clear mind -- capable of making better choices.

The awakened genius understands that there is no single change, thought or action that creates success, but rather a *synergistic* approach to life. There is no one of these health tips that is more important than the others. You will build balance when you work, play and rest in equal measure. Balance in the brain creates whole brain thinking--balance in diet and exercise produces a healthy, whole body.

Toxins in the body are the number one destroyer of good health. When you eat small portions of healthy food and drink plenty of pure water, you are keeping your body free from toxins. Toxicity will only build up in the body if you overeat regularly and if you consume toxic foods. You can also improve your chances of remaining toxin free by starting a regular exercise program. Now before you let that groan rumble up from your chest, read on.

Researchers have found that something very alarming happens as we grow older. At around the age of 27 our bodies begin a downward spiral. Women will average a loss of 2 pounds of muscle to fat a year; for men it is approximately 1 pound a year. If you are one of those people who says, "I just don't have time to exercise," think again. It doesn't take a genius to figure out that you don't have time not to! Have you noticed some of your older relatives getting smaller with age? The elderly in general are very small people. This is because the muscles that once held a strong healthy body in place have atrophied from lack of use.

There's no reason to let those know-it-all exercise fanatics tell you what to do or put you off. Create an exercise plan that *you* can enjoy. *A little bit of something is better than a whole lot of nothing*! Have you ever considered exercising but found yourself procrastinating because you don't have time to get serious about an exercise program? If so, realize that exercising doesn't have to be a "program," at all. Studies show that simply going for a walk is one of the best forms of exercise. You will receive great rewards from any exercise capable of stimulating the cardiovascular system. A simple rule of thumb is to start slow and build up to 22 - 45 minutes every other day. Of course you can exercise every day but remember, for the genius balance is what's most important. Make it fun. You have every right to enjoy good health -- it is your natural birthright.

MARTIAL ARTS -- A WHOLE BRAIN ACTIVITY

I have developed a personal interest in the martial arts. I was fascinated by an article that explained how Taekwondo training helps to stimulate whole brain activity. After completing my tape project for the American Taekwondo Association, I was so enthralled by my research results with some local black belts that I felt compelled to join in and find out for myself. I must say that my training has thus far exceeded my expectations. I find karate mentally and emotionally rewarding, and the physical flexibility allows my muscles and tendons to remain strong.

In researching the martial arts, I discovered that although there are many styles, they all believe in the foundation of balance and equilibrium in life. Numerous are the stories of troubled students entering into karate training and gaining the self-esteem and self-worth necessary to successfully complete their schooling. But it is important before joining any organization that you research the motivation of the instructors. I have found the American Taekwondo Association (ATA) and American Taekwondo Fellowship (ATF) to be among the leaders in the field. They are organizations sharing a commitment to integrity and excellence for the complete person.

The secret to the power of your body as outlined in all martial arts is in balance. I was quick to learn that if I throw a punch or kick without balance and without using *reaction force*[5], I am completely vulnerable to attack. But as I learned the secret of reaction force I gained a great sense of balance, even while performing the most difficult kicks. This same reaction force is occurring in your life right now. Some of you are reading this and thinking to yourselves, "I already have a healthy body; I'm young and strong." This is true today, but will it be true ten years from now?

Some of you might not be willing to sacrifice a few minutes out of your day in exchange for your strength. I ask you to think again, and to consider the path of least resistance. Remember it is easier to keep a large rock rolling than it is to start one in motion. As you begin your exercise program you might find it hard to motivate yourself at first. After a time, however, it will become a part of your nature.

I still remember the words of my wrestling coach who would stand, arms crossed and toe tapping, impatiently waiting after a wrestling match for me to finish my weight training so that he could lock up. Whenever he would tell me to hurry up, I would remind him of what he told us on our first day of practice: *"An athlete does what's expected; a champion does just a little bit more."* He would struggle to hold back his big, toothy grin, but finally he would give in and tell me to take all the time I needed. There might not be a

---

[5]**Reaction Force** is the act of using a reverse action to give punches and kicks in martial arts more force and power.

gold medal in your future, but tests have shown that exercise reduces your risk of all major diseases.

A very wise man (and my mentor) by the name of Dr. Gil Gilly once told me that hell is when the person you are today encounters the person you want to be and you know you will never meet that person on your present path of life. Heaven, on the other hand, is when the person you are today meets the person you are going to be and they come together like old friends, knowing they will one day meet on the road of life. The trophy will be yours when you can look at your body in the mirror and enjoy knowing that you have the privilege of using that body in health and harmony for another day.

DOCTOR SUNSHINE

One of the greatest medicines available is 100% natural and absolutely free! Of all the healing properties obtainable today, *sunshine* is probably the least utilized and most misunderstood. People are made to fear the sun as an alleged cause of cancer. Dr. Jay Hoffman, in his book ***The Missing Link in the Medical Curriculum***[6] tells us, "Many people stay away from sunlight from fear of developing skin cancer. This is most unfortunate, because sunlight is one of the greatest healers. If your bloodstream is clean, alkaline, free-flowing, and plenty of oxygen is getting through to your tissues, you will have a healthy body, which includes healthy skin. Your tissues must have 45% oxygen at all times ...If an individual is living on a low fat diet he can get all the sunshine he wants and it will do him good."

Dr. Zane Kime[7] wrote an entire book on the topic of sunshine, "If the sun is able to eliminate bacteria from our air, water and skin, and is able to strengthen the immune system of a host," writes Dr. Kime, "it would naturally follow that persons regularly exposed to ultraviolet light would develop fewer illnesses. This is exactly the clinical result observed."

---

[6]The Missing Link in the Medical Curriculum, which is Food Chemistry in its Relationship to Body Chemistry, Dr. Jay Hoffman, Professional Press Publishing Company, Valley Center, California, 1981, 1982, 1984

[7]Zane R. Kime, M.D., M.S., Sunlight Could Save Your Life, World Health Publications, Penryn, CA 95663, 1980.

The sun means life itself--it is the source of all energy and is absolutely essential for anything to live or grow. Dr. Hoffman recommends regular sunbathing of 20 to 30 minutes per day, depending upon your skin type, and then following up with an invigorating shower to cleanse away all of the toxins that the sun has drawn out.

If you enjoy a healthy, low-fat diet and get plenty of exercise, as I described earlier, there is no reason to fear the sunshine. When you go outdoors and enjoy the fresh air and sunshine, you are improving your health, both physically and mentally, and your longevity.

SLEEP = REJUVENATION

We live in a busy, busy world -- so busy that millions of people do not get enough sleep. Then to overcome their fatigue they drink caffeinated beverages for that little "pick-me-up," thus forcing the body to work beyond its capacity, creating exhaustion, and so on. A vicious little cycle, indeed; one that can destroy health. We are all subject to certain laws of nature and you cannot expect to have good health if you ignore them. After work or exercise you must have rest and deep sleep for rejuvenation.

The genius knows that sleep is a time for healing the body and refreshing the mind. Geniuses learn how to regulate their daily activities to allow time for rest and to avoid a schedule that leaves them feeling exhausted and overwhelmed. They know that rushing around from one place to the next creates exhaustion and inhibits success. The brilliant mind is ever cognizant of what is achieved during deep, peaceful sleep.

The genius puts the body to bed and then plans for the optimum benefits of a deep, rejuvenating sleep. To reap the rewards of a good night's sleep, the genius goes to sleep with an empty stomach. It is true that eating a large meal will make you sleepy at first, but that full stomach may later disturb your sleep. After a filling meal you often feel sleepy because the blood is being diverted from your brain to the stomach. Later, however, when it is time for deep sleep, you may become restless and uncomfortable because the digestive process has slowed or shut down even when your stomach is full. Going to bed on an empty stomach will usually allow you a deep, peaceful sleep so that only seven or eight hours are required. Going to bed after a hearty fill, however, may extend that requirement to ten hours or more. Geniuses know they will sleep best with a stomach that is nearly empty.

Deepak Chopra in his book **Quantum Healing**[8], carefully describes the indescribable: " ...the body is awakened in the morning not by a rude internal alarm but by a series of timed signals, at first mild, then progressively stronger, that lift us from deep sleep by stages." For this reason, if you have not received your full amount of sleep and are awakened by the ring of an alarm, your brain and body will awaken with a grogginess that may stay with you throughout the course of your day.

So, you might ask, how do I know how much sleep I need? How do I go about awakening on my own each morning when I have to be to school/work at 8:00 a.m.?

The amount of sleep required varies from person to person. Some people can get by on as little as five hours of sleep, whereas many need nine or more. One way to determine your sleep requirements is to plan a few days when you do not need to arise at a certain time. Go to bed early on those nights and sleep until you awaken naturally; then count the number of hours you slept. Do this for several nights so that by the second or third night you have caught up on your sleep and will know the true number of hours required. Once you have determined the number of hours you need, you can plan to put yourself to bed each night so as to awaken when you need to, feeling refreshed and energetic for the day's activities.

As an awakening genius, you will want to get to know your body well -- its rhythms, patterns and needs. Remember, it is through your body that you express yourself. The combination of a healthy diet, exercise, rest and recreation will create a healthy and balanced you -- now that's sheer genius!

---

[8]Chopra, Deepak, M.D., Quantum Healing
Bantam, Doubleday, Dell Publishing Group, Inc., 1989/1990.

CHAPTER NINE

THE BODY-GENIUS

I still remember that hot August morning; getting up at daybreak ready for a week of training at the University of Michigan's football camp. I had spent nearly the entire summer painting our house to earn the money for registration. At times some of my childhood buddies would hang around at the foot of the ladder, chomping on grass and trying to convince me that I was wasting my time and money. "Come on, Porter," their voices would echo, "you're already great at football; what good can a week do?"

Truthfully, I wasn't sure what to expect either, but all my doubts were erased shortly after I arrived on that first day. By mid-morning it was sweltering, but I paid little attention to the heat. As we were stretching, the track coach began teaching us a technique of visualization. He was having us imagine the running form that he would be demonstrating later that day. This is great, I thought to myself. Maybe my father wasn't so far off with all of his talk about using our minds. But I soon found out that however easy it was to do the moves in my mind, it would take some practice before my body would catch on.

In those first few days I mastered the running form, first with my mind and then with my body. A confidence built up within me that is impossible to describe. When I asked the coach about these new sensations, he answered with a broad smile, "Sounds like you found your *game face,* Porter. Good job."

This kind of confidence can only happen if your body and mind are in full agreement. I still remember that first game of my junior year when I looked over at my best friend, Jamie, an all-star defensive tackle whose athletic abilities I had always envied. I could hardly believe it; I was actually starting as outside linebacker. Even though I was only 5' 8" at the time, I ended the year as one of the top linebackers in the city and earned honorable mention for the conference. This might not seem terribly impressive to you, but as a freshman I was what I have come to call a "30-Second Wonder." All week long I would practice my heart out with the team. Then, on game day, when there were 30 seconds left to play, the coach would let me into the game. I would run out onto the field determined to show the coach what I was made of, only to find myself watching in wonder as the seconds ticked off the clock -- wondering what the heck I was doing playing this stupid sport anyway!

What changed my attitude? The realization that my mind and body were not working together. In my mind I wanted to be a football superstar, but as a freshman I was 5'2" and weighed only 112 pounds; my body hadn't yet agreed. Luck was with me, however, for I came from a large family and my "big" brother, Michael, had already been through the "I'm too small to play football" thing. Everyone made fun of him when he bought the Charles Atlas course, yet to this day he maintains a physique of which Charles himself would be proud. Michael and my father quickly advised me that if I truly was going to be captain of the football team, it would take much more than faith. I would need to do some work. I was ready to make that mind/body agreement happen.

I not only achieved the status of three-sport captain, but also received ten varsity letters before it was over. To this day I am amazed at how easy it all became after I made the choice to take a positive approach and back it up with physical action. One of the best benefits of this positive attitude was the fact that I also maintained an honor roll status through high school and into college.

*"The doctor of the future will give no medicine,
but will interest his patients in the care of the human
frame, in diet, and in the cause
and prevention of disease."*

THOMAS A. EDISON

FIGHT OR FLIGHT

What role, if any, does the physical body play in programming genius behaviors or in the running of old ones for you? A big one.

Your body has been trained by your mind to react to situations in life. This is described in most literature as the "Fight or Flight Syndrome."

At times we are all faced with situations that seem too big or too complex to handle. Most people will respond by either fighting -- doing everything possible to change the situation, or by running from the situation and taking an easy way out -- such as drugs or alcohol.

But, for geniuses, there are always other options! Geniuses know that a more viable solution is just around the corner and within reach. These are the people who know that situations are not always what they seem. They are ever aware that the predicaments in which they find themselves are based on their perception of reality, not reality itself. If it's time for a new perception or a different response, the genius mind will make note of the current response to the situation, whether it be fear, frustration or anxiety, and then begin a process of visualization to create a new concept or view and a more effective way to respond.

THE TIME MACHINE

Let's say that you could step into a time machine and go back to the second grade. You are teleported through time, but you can take all the knowledge you have today along with you. How would you do? The work would probably seem quite simple to you. You might even start to wonder why the other kids are not bored to tears.

Now what if you took that same attitude into the future? You could take your time machine into the future and imagine that you have already been through an experience in which you responded appropriately. And then, just to be sure, you take your time machine back to see, hear and experience that situation a minimum of five different ways, so that when you arrive you are prepared for every possibility. It's all old hat to you now -- like going back to the second grade.

Can you imagine the kind of flexibility this would build? -- which, by the way, is another key to genius. Just imagine what resources this would create in the storehouse of your mind. You now have many new experiences from which to draw information because the mind won't know the difference between what is "real" or what you imagined. It is a well known truth that the same neurochannels will fire whether the scene is real or mentally imagined. This is a strong key to the genius and could explain why so many geniuses can go to sleep with a problem in mind and then find the next day offers the precise solution.

Could it be possible that our internal genius never sleeps, but if properly motivated continues to work at a level far beyond that of our conscious mind or understanding? It has been my conclusion that this most definitely is the case.

It was an incredibly warm March night in Phoenix. Although the thermostat would often shoot past 100 degrees in early spring, my mother refused to turn on the air conditioning until what she called, "real summer." It was very late and I was making my way through the living room, stepping gingerly over the array of limp, moist bodies. My younger brothers and sisters were all sleeping soundly as the television set cast a glimmering sheen across the room. As I flipped off the television, my brother John, who was about twelve at the time, bolted upright and began speaking out loud.

I couldn't resist the temptation to see if I could get him to communicate with me while he was asleep. The results amazed me and surprised him even more.

"John," I asked, "what are you doing?"

"My math," he promptly answered in a matter of fact voice.

I was astonished by his reply. This was a Saturday night and he had been sleeping in front of the television for several hours. There was a pad of paper and pen on the coffee table so I grabbed it and began making notes.

"What problems are you working on?" I asked.

He gave me a series of answers which made no sense at the time. "John," I asked, "is there anyone there helping you to learn your math?"

"Yes," he said. "He's seven feet tall and has green skin," he hesitated, then added, "and he's wearing a silver space suit."

John had always been an avid reader of science fiction, so when I asked who was helping him, he gave me some futuristic space name. At many of our family gatherings, John's "giant space teacher," became the topic of many jokes.

The most astounding revelation in John's sleep-talk came when we discovered all the answers John had given me in his sleep were solutions to problems in the next chapter of his math textbook. Later, when John and I discussed what had happened, he told me that he didn't consciously know he was studying ahead. "It just comes to me easily," he said.

John has since graduated number four in his high school class with a grade average above four points because he had completed several college courses before finishing high school. John was fortunate, for his mind was instilled with all the techniques in this book from the very moment he was born. He was supported in his studies and encouraged to succeed. His genius mind was simply using that sleep time to prepare him for another success.

Albert Einstein devised his own unique methods of accelerated learning. He tended to depend on his own brain to bring him answers to questions or solutions to problems. One such learning process was used during his "cat naps." He would work diligently until he came to a stuck point. He would then go to sleep giving his mind the suggestion to continue thinking of options so upon awakening he could continue his work in a successful direction. He was using his genius to solve problems even while deeply asleep.

Before going to bed at night, take a moment to write down any questions to which you would like the answers or any problems for which you need a solution. At first it will be important to write the questions or problems down to program what you are seeking into the *other-than-conscious* mind. This process will quickly become a habit, and there will no longer be a need to write them down.

Just imagine how your life can change if you simply begin directing your *other-than-conscious* mind to bring you dreams that result in success programs, answers and solutions. You would literally be experiencing *quantum learning.*

"Creativity can solve
almost any problem.
The creative act,
the defeat of habit
by originality,
overcomes everything.

George Lois

"The quantum revolution made it inevitable that our world view would change. Quantum physics proved that the infinite variety of objects we see around us - stars, galaxies, mountains, trees, butterflies, and amoebas - is connected by infinite, eternal, unbound quantum fields, a kind of invisible quilt that has all of creation stitched into it. Objects that look separate and distinct to us in fact are all sewn into the design of this vast quilt. The hard edges of any object, such as a chair or table, are an illusion forced upon us by the limits of our sight. If we had eyes tuned to the quantum world, we would see these edges blur and finally melt, giving way to unlimited quantum fields."

Deepak Chopra, M.D.
Perfect Health

CHAPTER TEN

QUANTUM LEARNING

Are you ready to take the next evolutionary step? Are you prepared to know the difference between all that you *have known* as truth and all that *will be* true for you? If so, then maybe like those pioneering souls in 1905 who looked at Einstein with confusion when first hearing his Theory of Relativity, you will begin to understand the relationship between the *small* changes you will make in the classroom and at home and the *big* difference it will make in awakening your genius.

The atom was once considered the smallest particle of matter. The word atom stems from the Greek, "not able to be divided." Today science tells us that an atom is actually made up of tiny bits of matter zooming around at lightening speed in a massive empty void. This vacant expanse has been compared to outer space. Relatively speaking, the span between two electrons is wider than the space between earth and sun. Suddenly we are thrust into *quantum reality*.

The word *quantum* stems from Latin and basically means, "how much?" A quantum of anything is its smallest unit of energy. In quantum physics this can mean light, electricity, or any other energy that can possibly exist -- including you! Many people find quantum physics incomprehensible because it completely defies everything we perceive as the "real" world. There is no solid matter in the quantum world. The skeptics ask, "How can this be? I can touch a table and I can reach out and touch you!" The truth is that the solidity of a table and even our human body is an illusion. It is the limitation of our senses of vision and touch that keeps us from knowing this quantum world.

The theory of Quantum Learning deals with the small, but probably most neglected, segment of learning -- *how* we learn. Great scholars have put thousands of hours, sweat and possibly a few tears, into developing the fine courses taught in our modern schools. Yet, it would seem that only a few pioneering teachers are attempting to truly solve the seemingly mysterious question -- Why do so many children (and adults) have so much trouble with learning?

This is no great mystery. The true problem resides in the reality that no one ever takes the time to motivate the student to learn. Quantum Learning is a set of learning patterns and techniques that build rapport with the inner genius -- the part of the mind that can and will absorb book work easily and readily, even instantly, if put into the right state.

Let's go on a little journey to a place called *understanding.* You are boarding NASA's latest model flying saucer and are readying for lift off. Imagine yourself going to visit an inhabited planet that was just discovered in another solar system. The solar system is in a distant galaxy and is part of a space-time continuum totally different from ours on Earth. No one knows your language or customs and you know nothing of theirs.

Upon entering the new galaxy, you experience a sudden and brilliant flash of light. At that moment you realize that you no longer have use of your eyes, ears or any part of your body. Indeed, your body suddenly seems very strange, not uncomfortable really, just different, and you wonder why you no longer feel like yourself.

When the NASA saucer gently lands on the planet, you are carried off the ship by two beings whose sole job is to help you cope and learn in your new environment. These two helpers are highly qualified. They have already been through much of what you will be experiencing.

You were never informed that anything strange would happen, but when you arrive on this distant world you instantly forget everything about where you came from and you remember nothing of your mission. All you know is what your guides tell you. Settling into this new life proves rather easy because your guides are very caring and shower love and attention upon you.

With time you learn to see through your new eyes and begin to categorize and memorize the new environment that surrounds you. You begin to notice that on this planet they use their mouths to utter sounds and the other beings seem to somehow understand what it all means. So again you start the process of categorizing and memorizing, this time the sounds and their meaning. Sometimes you are right and other times you are wrong, but with time you are able to make sense out of all the noises and even learn to utter a few words of your own. Then one day you notice that when you make the sounds your two guides become very happy. You decide to learn more and become better with the sounds every day. When your guides are happy, it somehow makes you feel happy too.

At the same time you are discovering the world, you also are learning to use the new-fangled body that you acquired upon entering the solar system. You look out with your new eyes and notice that everyone else can easily walk around on their three feet and accomplish many tasks with their four arms. Slowly you begin to experiment with your new body.

Your guides have now decided it is time to allow you to feed yourself. Even though you make a mess and feel totally unprepared for an existence with four arms, you soon learn how to maneuver your food up to one of your mouths. You have two mouths, by the way, and find it easy to make sounds with both.

Just when you start to figure out the routine, your guides decide it is time to send you off to school to learn. Perhaps you are now feeling sorry for yourself and totally confused. You have only been here a few short years and now you are being asked to understand an entirely new culture! The guides are there to reassure you and, since they seem so excited about it all, you decide to play along. When you arrive at the school something wonderful happens -- you find out there are lots of others just like you. They have all been through what you have; they learned to use their bodies, and to make sounds in much the same way that you have. You are all given a guide and they call this one "teacher."

At first it's great fun. You get to play games and run around outdoors under the warmth of two suns, but one day "teacher" hands you a book. When you look down at the pages all you see is "🕯⌨♐●♑◆⌨♐&❑📪♌". You have never seen anything like this before! Then "teacher" announces that you will be given only a few short months to learn this written language. Suddenly life on this new world seems so unfair ... .

On this planet, a place that is vastly different from the world in which you lived on Earth, your problems are just beginning, and will probably get worse and worse if you continue to think of life as difficult and "unfair." Let's stop here. I want to take you to a real place and a real time.

When you were born you came into a completely new world. You had no information available to help you learn or understand. If it wasn't for your parents you wouldn't have made it for even a week. Given the quantity of information you had to learn as a child, the fact that you learned anything at all is truly miraculous. It is that continuous miracle that I speak of when I say you are awakening your genius. Even the agnostics[9] have to agree that something far greater than their conscious mind is keeping their thoughts in order. Whatever this master librarian is, it is nothing short of ***super***conscious. So how do we tap into this quantum amount of information? Simply follow the steps outlined in this chapter.

GOING TO THE ALPHA STATE

As was explained in Chapter 4, the purpose of Alpha is to get the brain into a ***positive*** and ***relaxed*** state. It has been proven that geniuses, such as Einstein, function primarily in the Alpha state. Einstein had an uncanny ability to function on small amounts of sleep. He would take what he called "cat naps," sleeping for ten to fifteen minutes six or seven times a day. It has since been discovered that the brain is unable to discern time at the Alpha level of the mind. Einstein was able to convince his brain and body that this was all the sleep he needed.

During World War II fighter pilots were specially trained to get by on very little sleep. They were given a spoon and metal pan and told that the body only needed as much sleep as it would take to lose their grip on the spoon. When the spoon fell it would hit the pan and the loud noise awakened them. These are but a few examples of how the Alpha state has been utilized in the past.

Another benefit of the Alpha state is the heightened level of ***creativity*** that is produced in a relaxed state. It was found through the study of music that the baroque style of classical music is very conducive to relaxation, which in turn brings about the generation of Alpha brain wave patterns. It has also been found that people respond with greater comprehension when the material is being assimilated with soft relaxing music. The authors of the book

---

[9]In this context, agnostics are defined as people who believe that the origins of our world and existence itself is beyond knowing or understanding by humankind.

Superlearning[10] tell us that if people can be trained to learn in the Alpha state through the use of music their retention will increase dramatically. Many tests have been performed on this theory and it has consistently proven to be so.

The next time you are in the doctor or dentist's office notice the type of music playing in the background. It is almost always a soothing, relaxing instrumental type of music. Medical professionals know the benefit in keeping their patients relaxed. It is in the Alpha and Theta states where fear and frustration are left behind and feelings of peace and rejuvenation take their place. Being out in nature is also a good way to invite the mind to go to the level known as Alpha.

You might have wondered how as a child you were able to learn such a multitude of behaviors and attitudes so effortlessly. Researchers have found that children maintain a strong Alpha brain wave pattern and display very little Beta. Perhaps this is why it is so often said that children are like sponges absorbing everything without judgment.

One typical sunny Phoenix day Tanya, our receptionist, stepped into my office with the most puzzled look on her face. She handed me a message slip and then just stood there. Her eyes darted up and down, then side to side as she searched for the best way to explain the note she had just placed in my hands. I gazed at the message wondering what the confusion was all about. It seemed clear: someone named "Ricardo" called and wanted to know if I could help him learn a language. Seemed like usual stuff to me. Then Tanya found her voice.

"This client called ...his name's Ricardo ...he asked if we could help him to learn a language," she said. "But, when I told him yes, we have helped people to integrate new languages very quickly, he tells me that he already knows the language! So I asked Ricardo what he wants us to do for him and he says help him to learn a language." She hesitated. "So I said, let me have the doctor call you back."

"So there you go," she said, patting the slip in my hand. She rolled her eyes then turned on her heel and walked back to her

---

[10]Ostrander, Sheila and Schroeder, Lynn, Superlearning, Delacort Press, New York, NY, 1979

desk. It seemed Tanya was convinced this guy was some kind of a nut.

I lifted the receiver and dialed the number on the slip. A man with a pleasant Spanish accent answered, "Hello, may I help you?"

I asked for Ricardo and almost immediately another male voice with a mild Spanish inflection came on the line. "This is Ricardo." he said.

I listened carefully as Ricardo unraveled his unusual story. Ricardo was a singer. He was very famous in the Spanish speaking countries and could sing Spanish perfectly. However, in the process of growing up in English-speaking Phoenix, Arizona, he had lost all knowledge of conversational Spanish. Ricardo's predicament intrigued me so I agreed to meet with him at his home later that week.

When I arrived I followed a long sloping driveway around to the back of an exquisite stucco house. My shoes crunched on the tiny pebbled walkway leading through a neatly manicured garden and to the entrance of his recording studio. I was quite impressed by the tasteful design of his home and gardens, so was not surprised to be greeted by a friendly, bustling housekeeper who guided me into the studio.

Ricardo's recording studio was the reason we chose to meet at his home rather than in my office. As I glanced around the room, I immediately knew that it had everything we would need to create the experience I was planning.

Because of the nature of the learning block, I felt it necessary to ask Ricardo just how it came to be that he was able to sing but not speak Spanish. Ricardo responded to my question in a straightforward, uninhibited manner. "When I was a young boy my family moved from Mexico City to Phoenix. When I started going to school all of the other kids made fun of me for speaking Spanish. I don't really know why, but many of them were relentless in their teasing. More times than I can remember these childish games would end up in fights."

As I listened to Ricardo's story, and the way he spoke, I would never have believed that he could sing so beautifully in Spanish. He had done a fine job of concealing his accent. The Spanish intonation would bleed through only occasionally and on certain words. As we discussed his difficulties, several of his comments confirmed his conviction that speaking Spanish as a child had made him appear stupid.

I asked Ricardo if he would allow me to use some relaxation processes with him. I told Ricardo about the Alpha and Theta brainwave patterns and explained that of my clients who stutter over 75% can speak perfectly while in a deeply relaxed state. Ricardo agreed and what happened next was astounding. During the state of relaxation he was able to remember the experience that prompted him to give up his heritage; to sing Spanish but never speak it.

His family had been very much involved with the church and its choir. The services they attended were spoken in Spanish and all the songs were sung in Spanish as well. Ricardo knew at a very early age that he could sing with the best of them. When he sang people showed him respect and praised his inspiring, melodious voice.

He now remembered one particular day when he walked home from choir practice feeling elated. He had made a personal vow: to master English and only sing in Spanish. This choice, he was sure, would solve all of his problems and it did indeed work for a time. He was old enough that he could now spend less and less time at home and more time with his English speaking buddies. He felt accepted as he learned to speak just like his friends, but could sing Spanish to his heart's content at the church.

As Ricardo grew older his singing career blossomed within the Spanish-speaking communities. Most of his American friends had gone separate ways and he was being thrust, more and more, into the public eye of the Spanish world. He soon found that he had created quite a dilemma for himself. He could understand Spanish, and he could sing it with fervor, but he couldn't speak it well at all.

I was called in at this particular time because Ricardo was scheduled for a long Mexican tour and had been invited to be a guest on several radio stations. His predicament had come full circle. He now believed that he would sound ignorant to his Spanish-speaking peers. It is no coincidence that Ricardo felt this way. After all, the law of mind is the law of belief. His childhood experiences had convinced him that speaking Spanish meant pain.

When I later played back the recording of our first session, Ricardo heard himself speaking fluent Spanish while in the relaxed and comfortable state of Alpha. While Ricardo was amazed to hear the sound of his own voice clearly speaking his native tongue, it didn't surprise me at all because I knew that the relaxed state of Alpha is the optimum learning place. One of the biggest blocks a student can experience when learning something new comes when past fear-related experiences freeze the natural ability to flow with the experience.

I had several sessions with Ricardo after his first success. Within a few short weeks he was able to speak flowing, natural Spanish while wide awake and alert. I was happy to hear that his tour was a success, not only on stage but also in interviews. He later commented that before starting the radio shows he would mentally repeat the words "calm and relaxed" (in Spanish, of course) and would feel himself enter into his relaxed state. Now he was able to speak freely and even have fun with the interviewers. Ricardo had placed himself into his *optimum learning state,* which you will learn to do next.

Everyone has the potential to learn anything -- from the ABC's to a foreign language; from two plus two to a new computer system. My father had a great little story which he frequently used in seminars to drive his point home.

"Fred" had a son named "Bill" who was taking a Spanish class in school. During a parent/teacher conference the Spanish instructor informed Fred that he didn't feel Bill was smart enough to learn Spanish. Fred turned to leave the conference feeling sad for his son. Then his steps halted, he thought for a moment, and turned back toward the teacher. "I think Bill will be all right," Fred said, standing tall and gazing straight into the instructor's eyes, "just teach him the same Spanish the dumb kids learn in Spain."

Fred's point was clearly made. Learning isn't about being "smart enough," but it is about being motivated enough. The best way to get motivated is to build a state for learning.

Four Steps To Your "Optimum Learning State":

Start out by playing relaxing music *(Baroque/ Melodic/ Classical/ New Age).* This starts the Alpha brain wave response and opens your mind to learning. This can be likened to the way you easily learned your ABC's when they were put into memory with a musical melody. The same will hold true for any learned information.

1. **Get into a comfortable position.** *Sit in a straight back chair, feet flat on the floor, eyes closed and rolled slightly upwards.*

2. **Use your imagination to create a place out in nature.** *This will be your* ***personal place of relaxation****. Make the experience as rich as possible. Imagine what you would see, hear and experience in this perfect place. Imagine that your personal place of relaxation has a golden dome around it and that in this place you can be or do anything you want. Remember to build yourself a shelter here in this place. You may improve or change anything you want.*

3. **Using your imagination, move through the successful fulfillment of all your goals and outcomes.** *Daily outcomes, weekly outcomes, monthly goals, yearly goals, etc. Imagine what it will be like when you accomplish these positive outcomes.* *(See Chapter 12)* *Again make the experience as rich as possible.*

4. **Bring yourself back by saying the words,** *"Wide awake ...wide awake."*

The whole process takes approximately 5-10 minutes. Take the time to do this **once a day**. It will then be easy for you to remember the state and return to it at any time ...all you will need to do is think about it and you will be there!

ENGAGING CONCENTRATION

How do *you* learn the best? The key factor in concentration is *interest* in the subject at hand. There is a simple technique for engaging your ability to concentrate.

Using the space below list five to ten things that you have learned in the past, that you enjoyed learning, and that you now do well.

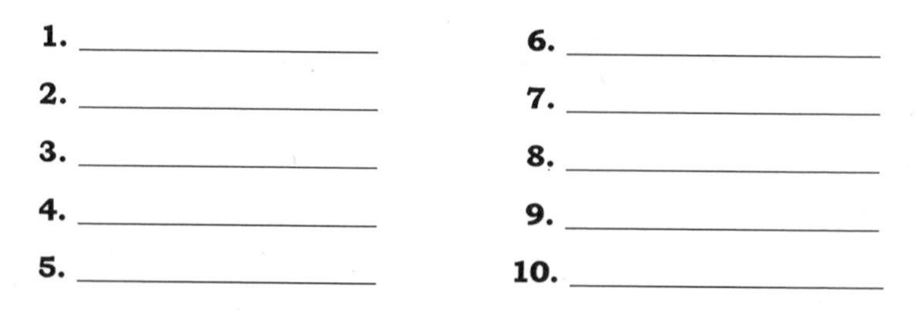

1. ____________________ **6.** ____________________

2. ____________________ **7.** ____________________

3. ____________________ **8.** ____________________

4. ____________________ **9.** ____________________

5. ____________________ **10.** ____________________

Take a moment to remember what it was like to learn each of these skills. What was it like internally while you were learning these skills and abilities of the past; the ones that you now can do so well?

What do you think would happen if you demonstrated the same concentration and interest that you had while learning an athletic activity, or a hobby, or in your favorite class, say typing, but during math or English? I guarantee you that your grades would improve. What if you could walk into your place of work on Monday morning with the same expectation for self-improvement you had for your favorite sport or pastime on Sunday?

Take a moment to recall the techniques of modeling (review Chapter 5). You can use these techniques to model yourself! The next time you find yourself losing interest, you will remember what it was like when you learned to shoot that jump shot, or the day you were given your own typewriter, or when you first discovered your favorite hobby (refer to your list above). You will remember what you were thinking and how your interest was piqued. Do you remember what colors were in the room? Did you say anything to yourself to get psyched up? Some coaches call this technique of modeling yourself setting a "zone" or a "game face." Although some students seem to settle into this magical state all by themselves, most of us must learn how to achieve it.

Without concentration the mind stores information as backwash. When you lack concentration it becomes easy for the conscious mind to drift off to other activities. If you haven't yet developed the skills of concentration, then this technique of modeling yourself is a simple and effective place to start. Who knows you better than you? In Chapter 11 you will learn the Three Finger Technique, a tool to help you focus your concentration and mental energy.

ADDITIONAL QUESTIONS FOR MODELING YOURSELF:

1. ***What were you thinking to yourself when you were learning something that you enjoyed in the past?***
2. ***How were you breathing?***
3. ***Was there a sense of excitement?***
4. ***At the time did you think you would ever learn to do it?***
5. ***Who was instructing you ...or were you alone?***
6. ***If you had a guide, what was your attitude about this person?***
7. ***Remember to the best of your ability what you were seeing, hearing and experiencing.***

CURIOSITY KILLED THE CAT ... ...BUT SATISFACTION BROUGHT IT BACK!

Hopefully you will easily discover, like those attending the Awaken The Genius Seminar, that learning happens naturally and effortlessly when your mind is curious. Simply watch the curiosity of a small child ...and then think about the vast quantities of information that same child learns in just a few short years. Imagine what it would be like if you had the curiosity of a child in every learning situation ...by all rights, you would have to be declared genius!

If a person is a **visual learner,** the most appropriate method of building curiosity will be through pictures and graphics that are pleasing to the sight. Think of those genius minds in the advertising industry. How do they pique our curiosity? Most of the time it is done with taste, but they will often use unrelated overtones that grab our attention. Beautiful women and men fill the screen promoting ugly cars and terrible smelling perfumes. Yet the products sell. Why? Because those unsuspecting buyers were curious and some even gullible enough to believe if they drive that car or use that particular scent, they will become like the model promoting it.

For the **auditory learners,** stimulating music and unique or interesting sounds will keep their interest and curiosity. Those cunning commercial producers frequently use catchy jingles and melodies to keep us in tune with their product. How many times have you found yourself humming the jingles you hear on television and radio?

As for the **kinesthetic learners** (the feelers), the best way to keep their interest and curiosity alive is to give them something to do with their hands. They need challenges that will allow them to use their emotions, or a "hunch" they may feel, or their mechanical ability.

"Do you remember the times of your life?" Our "good buddies" at the ad agencies frequently create their commercials to inspire an emotional response. When you read the line above, do you remember the Kodak commercials? I have seen my sister on the verge of tears while watching a television commercial. She is a kinesthetic learner.

Most teachers, you understand, are in a "catch twenty-two" situation. They need to keep their students interested, yet they are given all the different styles of learners mixed together. So how do we develop the win/win that everyone wants. Well, that is just what this book is all about. No one is *just* visual or auditory or kinesthetic. We all use every one of our senses every day. It is just that one mode usually becomes dominant and more comfortable than the others.

How To Break Through The Old Patterns

The best way to break out of this complacency is to take the risk, experiment and discover the greater parts of ourselves. That internal genius is starving for fresh new horizons ...it looks forward to climbing new mountains. If during the testing in Chapter 2 you discovered that your preference was visual, kinesthetic or auditory, then the point here is to find new ways to stimulate the other senses. Take time out to improve yourself.

If your visual score was highest, perhaps instead of going home and turning on the television, you will choose one evening a week to play relaxing music on the stereo and exercise your auditory senses. If you had a high auditory score, maybe a trip to an art gallery is in order. What about a high kinesthetic score? Consider getting dressed up and taking a trip to the local mall for a few hours of "people watching." These are only a few examples. You can use your creative genius to come up with ideas of your own. Consider asking those people who seem "different" to you about what they find exciting, and then go play.

When I went through phases of feeling rebellious, refusing to do homework or improve myself, my father, in all his wisdom, would remind me of the benefits of further education. He would look me square in the eye and say, "Patrick, you will always hear people complain about the cost of education, but no one ever wants to discuss the expense of ignorance."

This statement made a great impression on me and I have continually sought new information, techniques and training to improve my personal state and the success of my clients.

After my Dad's career as an alcoholic ended at 38 years of age, he decided to go back to school and earn his degree. He found his classes to be easy and enjoyable even after so many years away from the books. He said it was because his motivation had now been accelerated -- he was determined to work his way out of the factory environment and create a better life for himself and his family.

All was going well with his return to academia. Actually, he was in his glory. My Dad never minded letting everyone know about his accomplishments. He proudly boasted when he received straight A's during his first semester and was determined to repeat the performance the next semester as well. Then he met the professor. You might have experienced an instructor like this one when you were in school. The professor strode with great seriousness into the classroom, stopped suddenly, and gazed around the room peering at each face through hooded lids. "No one," he announced proudly, "I repeat ...no one ...ever gets an A in my class."

He proceeded to strut like a proud peacock in front of the class. A hush fell over the group. The professor's eyes shone like jewels, he had gotten just the reaction he'd expected. The students glanced at each other with apprehension. They knew it was going to be a rough semester.

I don't know what made my Dad do it, maybe his long battle with alcohol made him bolder than most, but upon hearing this pronouncement, he immediately raised his hand. The teacher's eyes shifted, reflecting his surprise. Clearly, he did not expect any questions. "Yes?" he asked.

"Does that mean you are not capable of teaching an A?" My father asked calmly and in his best matter-of-fact voice.

The room was totally silent. The professor became visibly uncomfortable. He had never thought of his mind set in that way. When there was no response forthcoming, Dad again spoke up, "Because if you are capable of teaching an A, then I am capable of earning one."

Dad did indeed earn his A that semester in this professor's class and in his other studies as well.

Have you ever met a teacher like this one? In my experience with school systems it is not an uncommon trait. I guess I must have inherited some of my father's boldness, because I will sometimes relay this story to a room full of teachers. At first they are caught a little off guard, but after they realize the importance of the story's message and the spirit with which it is presented, they relax.

First of all, if you are the teacher, you had better be smarter than your students. You have the test and the answers! Second, I would think the greatest success for educators in any field would be to know they have helped the students to become as skilled as or even better than they are. I believe that teachers should receive a report card at the same time as the student.

My dad and I have often taken training seminars together. He used to tell people that I was much smarter than he. He explained that this was because he taught me everything he knew and I also knew everything that I had learned on my own. Although at the time I knew this was not quite true, it was his way of building my self-confidence and self-esteem. The job of teacher or trainer should be considered a sacred trust. There is no room for inflated egos when the future is literally resting in their hands. I believe the good teachers are probably not paid near what they deserve.

If you are in a position of "power," whether a teacher, coach, boss or even a parent, you are in a position to dictate how others will learn and react around you. Therefore, these positions allow you to activate others to use their full potential. A true genius is very much like a true leader -- he or she leads by example.

THE CRAB SYNDROME

When a fisherman is gathering crabs he will find it necessary to keep a lid over the bucket for only as long as there is but one crab in the bucket. After the fisherman has captured two or more crabs he can leave the bucket open and unattended and the crabs will never escape. This is because every time one crab gets close to crawling out of the bucket one of the others will reach up with it's mighty claw and tug it back down.

Every time my Dad tried to stop the alcohol habit his old drinking buddies would come around trying to lure him back to his old ways. These men were like the crabs. If my father succeeded in stopping drinking, they might have to do it too. So continually they tried their darndest to drag him back down into the bucket with them.

You might know people like this too. Every time you start feeling good or it looks like you might be getting ahead, they are the first to tell you, "It'll just get worse again," or "it won't last forever." These people are afraid; they are fearful of using their mind and of getting ahead in life. They are scared stiff; frightened and frozen into a cube of their own making. They are afraid that if you do better, they will somehow be less. The genius philosophy is much wiser: *The better you become, the greater I will be, for we all share in this adventure called life.*

THE HUNDREDTH MONKEY[11]

The Japanese monkey Macaca fuscata has been observed in the wild for a period of over 40 years. In 1952, on the island of Koshima, scientists were providing monkeys with sweet potatoes dropped in the sand. The monkeys liked the taste of the raw sweet potatoes, but they found the dirt quite unpleasant. An 18-month-old female named Imo found she could solve the problem by washing the potatoes in a nearby stream. She taught this trick to her mother. Her playmates also learned this new way and they taught their mothers too. This cultural innovation was gradually picked up by various monkeys during the course of the scientific experiment.

Between 1952 and 1958, all the young monkeys learned to wash the sandy sweet potatoes to make them more palatable. Only the adults who imitated their children learned this social improvement. The other adults kept eating the dirty sweet potatoes. Then something startling took place. In the autumn of 1958, a certain number of Koshima monkeys were washing sweet potatoes--the exact number is not known. Let us suppose that when the sun rose one morning there were 99 monkeys on Koshima Island who had learned to wash their sweet potatoes. Let's further suppose that later that morning, the hundredth monkey learned to wash potatoes. Then it happened! By that evening almost everyone in the tribe was washing sweet potatoes before eating them. The added energy of this hundredth monkey somehow created an ideological breakthrough! What was most surprising to these scientists was that the habit of washing sweet potatoes then spontaneously jumped over the sea. Colonies of monkeys on other islands as well as the mainland troop of monkeys at Takasakiyama began washing their sweet potatoes!

Thus, it would seem that when a certain critical number achieves a genius awareness, this new awareness may be communicated from mind to mind. Although the exact number may vary, the Hundredth Monkey Phenomenon means that when only a limited number of people are exposed to an innovation, it remains the property of these people. But there is a specific point at which if only one more person tunes in to a new awareness, a field is strengthened so that this awareness reaches almost everyone. That person could be you, right this very moment, making a difference

---

[11]Porter, Michael J., II, Individual Impact

that could shift life forever. Stop reading for just a moment and imagine a planet full of geniuses all working together to improve the living environment for each other. Is this the kind of world in which you would like to live?

Experimental psychologists later proved the Hundredth Monkey Phenomenon in rat water maze tests conducted over a period of 35 years, first in America, then in Scotland and finally in Australia. The results showed that the rats became better and better in escaping from the maze as time went on and that this increased ability to learn was then transmitted geographically. The rats in Scotland learned quicker than the original subjects in America, and those in Australia learned fastest of all. This increased ability to learn affected all rats of the same breed whether or not they were descended from trained parents. They demonstrated this transmission of acquired understanding beyond the "usual" concepts of learning. Again, today's geniuses realize that we are all in this together.

What does all this Hundredth Monkey Phenomenon mean to you as an awakening genius? Perhaps that you are not only awakening yourself, but also the inner genius of those around you! It has already been discussed that we are bombarded with thousands of times more information than the people who lived at the turn of the century. Yet, the onslaught of information seems to have little or no effect on us; or so we think. The reality is that our internal genius, that part of our mind that is *other-than-conscious*, has been protecting us, gently teaching us how to manage such massive amounts of information.

I find myself staring incredulously when I hear educators claim that today's children are less intelligent than their predecessors. This is an insult to every parent. Today's students are certainly not less intelligent. They are just intelligent in a different way and need to be re-engaged into the flow of education. It's time we all admit the truth: today's children find school dull and boring. Let's look at this another way. How well would a child of the 40's have handled a Nintendo game? Most of them would probably have given up, considering the whole thing impossible. Yet today's children seem to have been born with the ability to easily manipulate these video games. Many adults will spend hours struggling with the device, trying to get that controller or joystick to bend to their will. Yet I have seen children as young as three pick up a video controller and, as if on instinct, know just how to manipulate it. How can this be? Was there a magical instant when that "hundredth" child became a video master, making it possible for all children to now do it with ease?

Then Monday morning arrives when each of these video wizards must return to school where they will probably retain less than 25% of what is presented that day. What causes this poor rate of absorption? The biggest problem is that most of what is taught is not usable information. Also, many children are being raised with the belief that even if they do learn in school, there is no guarantee for their future. These beliefs have a lot to do with prejudices handed down from generation to generation. When the parents feel hopeless it is a powerful emotion that can be transmitted to the child. That same helplessness can manifest in the classroom. When students feel that what they are doing is worthless, the brain will store the information as useless.

The children of today are gifted with the skills of tomorrow, whether the "powers that be" want to believe this or not. The current group of geniuses who are attending today's schools are, in all actuality, creating the future. Through their attitudes and beliefs the destiny of this planet will spring forth. In today's world it would seem that the values of the textbook publishers are much different than those of the students. Can you imagine a school of the future that is filled with multimedia interactive computers teaching children in a full sensory experience just how they can help create the world they prefer to live in? Would the students be re-engaged in the process? I think so.

Children, just like every other living organism, grow at varying rates. Yet, in our current school system they are all expected to develop their minds at the same pace. This presents a gap in the genius potential. Each student is like a seed; when students go to school, they need to be nurtured. When a seed is planted in fertile soil and then watered regularly, it will germinate and then develop into its own unique pattern.

If you have ever driven through the Midwest in late summer, you have probably seen unending rows of corn fields. The giant stalks are deep green and topped off with fluffy wheat colored tassels. You can even pick out the stalks where those brilliant yellow ears are safely tucked within its protective folds. Every now and then, however, you will come to a section of field where the stalks appear withered and weak. Here the tassels droop, the stalks themselves are the color of pea soup and tiny, undeveloped ears stand out like growths. What happened? Something must have gone wrong in this section of field. Every single seed planted held the same potential. Yet there was too little, or too much, of something that kept these stalks of corn from growing to their mighty potential.

When a plant starts to grow, it will first produce its roots so that it can receive the nutrients and support needed for growing tall and strong. While all this growth is occurring, from the surface it may seem as though nothing is happening. The gardener may become very impatient waiting for the seeds to sprout. If the gardener now tries to force that plant to grow, by adding additional water or plant food, she may hinder its natural growth, or even destroy it completely. Just because the gardener couldn't view the growth from the surface, it didn't mean that the plant was somehow deficient. The truth is that the plant was in the midst of wondrous growth all on its own.

Have you ever been out hiking and discovered a wildflower growing right out of solid rock? Did you wonder how it was able to stay alive? If you could look under those rocks, follow the plant under the surface, you would probably uncover an incredibly elaborate root system. All on its own the sheer genius of this little flower developed a root system capable of maintaining just the right amount of nutrients to sustain its life.

What does this have to do with genius? Everything. You can liken each child's genius to the seed -- it is loaded with potential. When these children are nurtured, loved and given the time to grow at their own pace, they will, like the mighty stalks of corn, grow to their full potential. When we try to force learning, or to fit each child into a mold, we are hindering and perhaps even destroying growth. With love and support every child can and will develop into a genius in his/her own right. They might not be the next Albert Einstein or Alexander Graham Bell, but then again, they might.

Many of our world's greatest inventions and discoveries were by people who others considered of average or lesser intelligence. Remember, Albert Einstein failed his first college entrance exam! Yet, just like the laboratory rats who collectively made getting through the maze easier and easier, these geniuses went against the odds; they created the hundredth monkey effect, and helped humanity out of the maze.

Are you now living within the confines of the maze? Do you feel trapped by what others have told you? Does someone or something else do your thinking for you? Even if you consider yourself a grown "stalk," it's never too late to move out of the maze; to wake up and use that untapped 95% of your brain. With your creativity, you could be the "one" who sparks the rest of humanity.

"When enough of us are aware of something, all of us become aware of it...The appreciation and love we have for ourselves and others creates an expanded energy field that becomes a growing power in the world."

KEN KEYES

Circle of Power Integration

Prepare yourself to enter the optimum place of learning that exists within your own mind. Imagine a circle on the floor in front of you ...and in that circle are all the skills and all the abilities you would need to attain your goals ...In that circle are positive attitudes and behaviors that will improve your life and your world as early as today, instantly and automatically .... place in that circle your own, individual OPTIMUM LEARNING STATE . . .

Now think of it is a color ...and step forward into the circle and imagine that the color is surrounding you and filling you up ... just like the beam on Star Trek ... the new thoughts and feelings are there, THE OPTIMUM LEARNING STATE IS THERE ...breathe the way you would be breathing if you knew this was going to work ...if you knew that every cell, every system and every organ is quickly and easily changing to a positive new feeling WITH LEARNING ABILITIES AND STRATEGIES BUILT IN ...and when you are full of that feeling step out of the circle and look around the room ...Now think of something that is negative to you, an unpleasant feeling like anxiety or THE FRUSTRATION AND CONFUSION OF LEARNING SOMETHING NEW ...as soon as you feel that emotion step back into the Circle of Power and notice how the energy shifts and your feelings change ...notice that the energy in your Circle of Power has increased ...This is a new resource for you ...every time you use this resource, the Circle of Power WITH YOUR OPTIMUM LEARNING STATE, it will be easier and easier for you ...soon it will become a permanent resource for you just when you need it the most. You will find that each time you turn on a light switch you are turning on your Circle of Power ...each time that you step through a doorway you will be stepping into your Circle of Power ...so that it will be there for you WITH YOUR OPTIMUM LEARNING STATE ...just when you need it the most.

TIME = CONDITIONING AND BELIEF

Probably the biggest stumbling block we encounter when attempting to master something new is the reality that learning takes time. So what is this space or place we call "time" anyway? In truth time is not real nor is it tangible. *Time is a constant only to the perceiver.* Have you ever been doing something enjoyable and "time" seemed to fly? Perhaps you glanced down at your watch and couldn't believe how late it was. On the other hand, have you ever been involved in an activity that you truly disliked and found "time" to drag? Perhaps you glanced at your watch and groaned, unable to believe you still had several hours to go. If you have experienced time in this manner then you understand what is meant by the statement, *time is a constant only to the perceiver.*

Geniuses realize the truth of time and use it to their advantage. For instance, if you are doing something you dislike, the chances of your remembering that experience are next to nil. But think of something pleasing and exciting that happened to you in your childhood and it might be stored with all the senses intact.

How does the awakened genius use time? It's simple--with *spaced repetition.* In the "real world," where the clock on the wall continues to tick, you are given only so many opportunities to practice a new skill or ability. You have only so much time for studying. But for those geniuses who step outside of time, there is all the time they need. *If you give a busy person something to do, you can rest assured that it will get done.* We all have the same amount of time available each day (review Bank of Time in **Initiation** section). The people who use their quantum[12] time most effectively get more done and receive the greatest benefits. The following section was designed to teach you this process. When you go inside your mind, you go to a world of all time -- the world of your imagination.

---

[12]**Quantum** meaning the thoughts between the thoughts; the other-than-conscious process.

SPACED REPETITION: USING THE INNER SCREEN

Martial arts training is an excellent example of spaced repetition. When you first learn your new *form*[13] the trained instructor doesn't teach you the entire sequence all at once. He or she will break it down into chunks of four or five moves. Once those moves are memorized and mastered with the flow and cadence necessary, then four or five more moves are added. The process of learning the entire form for each belt occurs over time. The trainee learns each form in a series of chunks and then adds them together. Before long, the trainee can perform a series of over one hundred moves in one form--a process that would have been extremely difficult and frustrating to learn all at once.

Spaced repetition is used in all major sports. Whether it is football, volleyball or tennis, you must first learn the fundamentals before you can learn the complete game. What good would it do to know the entire game of tennis without knowing how to swing the racket?

How do you, the awakening genius, use spaced repetition? First, you must be able to get more time from your minute. This is where visualization comes into play. Several leading sports psychologists claim that five minutes of visualization is worth two hours of physical practice. If this is the case, anything you choose to learn can be mastered in a very accelerated way by using the inner screen of your mind.

This technique is best used at night right before going to bed. Think of something you would like to improve or change. Remember, your mind will bring you what you dwell upon, so think of what you want and forget about what you don't want. Your mind doesn't care what you want to succeed at specifically; it assumes that whatever you are thinking about is a goal. So when you find your mind dwelling on the negative, you can rest assured that the power of your mind is actively figuring out the best way to bring you more of the same! The genius mind understands the power of thought and is constantly monitoring thinking patterns to create a flow toward positive outcomes. It is natural to experience feelings of

---

[13]A series of moves that you are required to learn in order to move up the ranks and receive your next belt.

depression, sadness and anger. Geniuses don't deny these feelings; they simply plan on more appropriate behaviors in the future.

SELF-DISCOVERY: SPACED REPETITION

Each night, while entering into your sleeping state, practice focusing your attention on the possibility of success. Let your mind access the same mechanism commonly known as *"daydreaming."* Simply allow daydreams of success to flow into your sleep patterns. When you practice this technique you will be just like the professional athletes who use mental rehearsal. Their neurochannels don't know the difference between a real touchdown and one that is imagined. Therefore, five minutes of mental rehearsal can equal two hours of physical practice. Only you know what you would like to improve and when you will make it so. Let your daydreams and your night-time dreams create the place in your mind where you make it "real."

PRACTICE MAKES PERFECT!

Your previous conditioning has a tremendous impact on your current life. How were you treated as a child? Are the memories in your mind of a loving, positive voice or of a critical, negative one? This plays a major role in your attitude toward learning. You have learned every behavior and attitude from your parents, friends and the environment around you. The biggest advantage you have going for you is the flexibility of your mind and its ability to change!

Just like the athlete, when information is stored in memory, your mind doesn't know the difference between fact and fantasy. So if you want to change an existing behavior, change the way it is stored in your mind and you will change the way the past idea and concept affects your daily life. No one grew up in a perfect learning environment. But the following dialogue can give you a more secure sense of self, almost as if you had experienced a much happier, healthier and more nurturing learning atmosphere throughout your life.

SELF-HELP DIALOGUE: RESOURCE ORGANIZER

Prepare yourself for a journey into the imagination where everything is in perfect order. Close your eyes and get into a comfortable position. Become aware of my voice. Notice how quickly you now find yourself relaxing. This time, bring up in your imagination A TIME WHEN YOU HAD DIFFICULTY LEARNING SOMETHING. Put it in a picture frame in your mind and slowly take the color out of it. As you do, notice that it becomes black and white. Any sounds that you are hearing about that past time are getting further and further away as if they are coming from the moon ...you know they exist, but they are now coming from outside of your awareness. Any feelings that still remain ...just take a moment and imagine them melting away into the black and white picture.

As this happens, imagine that the part of the picture that is white is getting whiter and whiter and begins to fill the whole screen of space ... When that occurs, whatever you learned from the experience instantly comes to you ...either consciously or unconsciously ...

Right at that moment you begin to remember a TIME WHEN YOU ENJOYED LEARNING SOMETHING NEW ...YOU FOUND IT EXCITING ...place this picture into the frame ...and as you do add more color to the picture. Bring up the sounds of this EXCITING time. Imagine that you are there breathing the way you were breathing then ...PLEASED WITH YOURSELF and FULFILLED.

Imagine looking through the eyes of the past ...hearing with the ears and sensing and feeling with the body of the past. At just that moment the whole picture in your mind comes to life as if it is all occurring now ... and as you go through the experience you can use your mind ...your powerful mind, to help you go back into your past and use this technique of shifting colors, sounds and feelings in such a way that whatever happened in your past that may in any way stop you from awakening to your full potential ...OR STOP YOU FROM LEARNING WITH EXCITEMENT ...would be changed permanently ...changed in such a way as to free up your mind to enjoy your future now ...this very moment .... As you take this time now to go back in time and shift memories, placing your awareness on the positive things in life AND ON THE JOY YOU HAVE FOUND IN LEARNING NEW THINGS, you will find a new world opening for you ...a new, bright and compelling world ...a world that will allow you to express and be who you are ...unlimited. You have started a process of change ...change is the nature of all things. From the spring flowers, warm summers, the falling leaves to the blankets of snow ...change is the one thing you can count on.

Now begin to think of where in the future the changes you are making today will influence you ...some today ...some next week ...some into the months to come. Only you know exactly where all of these positive changes will be. Some you will know about consciously ...some you will know about unconsciously

...they will just happen for you ...(pause)...

When you feel that you have made the necessary changes in your past and placed the necessary behaviors in your future and you are convinced that the changes have been made once and for all ...positively ...you can return back into the room. And, when you are back you can open your eyes.

God is all-pervading, but yet we have some scientists who assert, "We have searched all outer space, we have looked for Him on the moon; no, He is nowhere to be found. He does not exist." They do not know what to seek and where, still, they have impudence to assert that it is not found ...To search for God with instruments in the laboratory is like trying to cure pain in the stomach by pouring drops into the eye! There is a special instrument for that purpose which the past masters in that science have developed and spoken about. Equip yourselves with a clear eye through detachment and love, sharpen your sense of discrimination so that it has no prejudice or predilection, then you can see God in you, around you, in all that you know and feel and are.

-- Baba

CHAPTER ELEVEN

THE STUDENT GENIUS

It was a crisp fall day and I felt as if I was being given a new lease on life. I had made it to my high school sophomore year, the coaches were beginning to recognize my athletic ability, and I was getting higher grades than I had ever dreamed possible. I probably couldn't have explained what had happened to me, but for the first time in my life, I was feeling positive about myself. With the changes I had made to my thinking, I was now finding all my Dad's "mind stuff" rather useful.

Having been an athlete, the best way I have found to relate this state of concentration is with sports terms. Even in the classroom, I had now found my *game face*; I was in my *zone*. When you enter into this learning zone, you are opening up your entire learning system. It has been proven that if you are totally present in the experience, you have a greater likelihood of recalling the information. To paraphrase Albert Einstein, knowledge isn't in collecting information, but knowing where and how to retrieve it.

This chapter is for the student or anyone who wants to succeed in a classroom setting. These techniques are specially designed so that you are placing the information into your memory banks for rapid and easy retrieval. Each of these methods is done with ease and will quickly become positive habits that will successfully guide you through any learning situation.

Where to Begin

In ***Psycho-Linguistics*** one of the first techniques we teach you involves rolling your eyes up to the position of ten o'clock. In other words, you are rolling your eyes up slightly as if looking at the ten position on the clock. Scientific studies have proven that this eye position will activate an Alpha brainwave response, which is considered the ideal learning state. In a classroom setting, focusing above the teacher's head and then defocusing the eyes will achieve the same results. It is not necessary to maintain this eye position; usually 30 to 60 seconds will do. If you later find your mind wandering, simply move your eyes into the Alpha position once again.

Tips for Taking Notes and Reading Assignments

Taking notes is probably the most effective method for reviewing information and committing it to memory. Unfortunately, note-taking is also the most ineffectively utilized technique. Studies have shown that less than 70% of the people who take notes ever review them.

Although our current system requires that students recall and memorize facts, true geniuses will form a way to organize their thoughts so that they retain the knowledge of *where to find the answers*. Even Einstein didn't possess a photographic memory. Many people who knew Albert Einstein thought him to be rather eccentric and absentminded. To him, however, everything he did had a purpose behind it. For example, he wore the same identical outfit every day. All of his shoes, shirts and suits were exactly alike. He explained this idiosyncrasy rather simply; he chose not to clutter his mind with useless information. Einstein could find no value in using up mind power remembering facts that could be recorded and then recalled later. He preferred to leave his mind open to creative avenues that could benefit humanity.

With this in mind review the note taking procedures below:

THE POWER OF X!

1. Always start new subjects on fresh paper.
2. Draw an "X" on the paper, the full size of the page, from corner to corner. Studies have shown that when you draw an "X" on your paper you build a visual bridge between the right and left hemisphere of the brain. You can fill in the page as you would normally, just write over top of the "X."
3. Write all notes as neatly as you can and abbreviate as often as possible.
4. Review your notes as follows. Draw an "X" on a blank sheet of paper. Using your memory and without looking at the original notes write as much as you can remember. Study only those items that you didn't write. *Continue this step until you remember everything you need for success.*

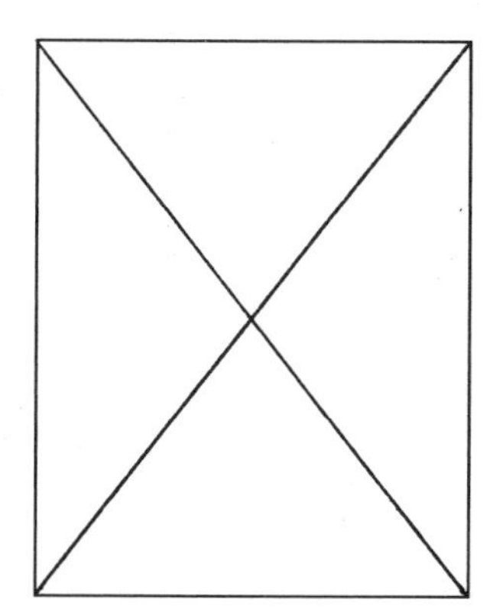

Four Steps To Improve Reading Speed And Comprehension

Improving Text Book Learning

1. Put yourself into your **optimum learning state**. (See page 157.)
2. Open the material with a fresh attitude and read first the **italicized** print, **bold** print and look at all the **pictures**.
3. Read through the headings and reformat them into questions in your mind. Read only the first sentence in each paragraph.
4. Skim through the bold print once again forming the statements into questions (you are building curiosity), then read through the material in each section only until you have an understanding of that section and have the answer to your question. It is important to stop reading when you begin to feel boredom or when you have found the answer.

IMPORTANT TIP - *Always stop, close the book and look around the room any time you feel yourself growing tired or losing interest. This ensures that your time spent reading is devoted to committing the information to memory.*

ANOTHER TIP - *Always remember to play soft and relaxing music in the background while reading. This will add to the experience and make it more positive. Your mind will make a habit of creating your optimum learning state whenever the soft music is heard.*

READING FICTION - *Use the above tips when appropriate. Most fiction is filled with description and detailed information. When you are reading for enjoyment, you will probably want to read every word, but if your mind starts feeling bogged down with uninteresting information your comprehension will suffer. Once you have trained your mind to function in the learning state, you may find your mind skimming over the information that has no bearing on the plot. Read fiction with the enjoyment of creating a movie in your mind.*

How To Improve Your Listening Skills

How well do you comprehend and learn during a lecture? Again, you can activate your learning ability through the use of your Circle of Power. Imagine that while activating the resource *(the Circle of Power)* you are remembering a time when you were totally interested in the information being presented.

Circle of Power for Listening Skills

You have activated your Circle of Power many times now. Before you step into it, guide yourself through the following:

Remember a time when you were listening to something and you thoroughly enjoyed it. This could be anything at all from your favorite music to a boyfriend or girlfriend speaking.

Remember to the best of your ability what your mental attitude was at that time. What was your mind thinking? Use your imagination to remember what you were seeing, hearing and experiencing. Make this as real as possible ...then step into your Circle of Power.

You will find that each time you turn on a light switch you are turning on your Circle of Power ...each time that you step through a doorway you will be stepping into your Circle of Power...so that it will be there for you WITH YOUR OPTIMUM LEARNING AND LISTENING STATES ...just when you need it the most.

THREE FINGER TECHNIQUE

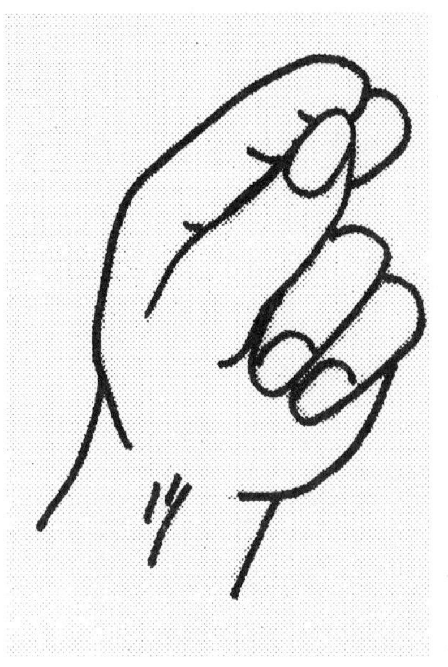

After setting up the Circle of Power start to associate that state into your listening state by using what is known as the Three Finger Technique.

This process is done by putting your index and middle fingers together with the thumb. (See example in illustration.) The main purpose of the Three Finger Technique is to trigger the mind into a listening mode. When using the Three Finger Technique you are queuing the mind on an internal level through an anchoring process[14] using the sense of touch. This technique can also be used in reading to improve concentration and comprehension.

Create the Circle of Power and when you have attained your optimum learning state, press your fingers together as shown in the illustration. Do this several times until just touching your fingers together brings about the desired feeling. Whenever you are in a learning environment, and want to be completely in tune with the speaker, simply activate your optimum learning state by pressing your fingers together. Try it and you may amaze yourself.

---

[14] Anchoring is the process of recalling a mental state with little or no conscious cues.

WALK A MILE IN YOUR TEACHERS' SHOES

What if you could see ***you*** through the eyes of your teachers? The next step in improving listening skills is to imagine that, during the lecture or class, you are the teacher or speaker.

Imagine that you could see through the speaker's eyes, hear through his or her ears and sense and feel with his or her body. Take a moment to gain an understanding of all that this presenter has to think about. Then using the space below imagine what you look like to him or her.

1. Do you look interested?
2. Do you ask questions?
3. Are you sitting up in your chair?
4. How would you describe yourself from your teacher's perspective?

As you awaken your genius, you will find that the moment you make this mental observation you will immediately perk up, becoming more interested. You will find yourself asking questions to clarify fuzzy areas of your thinking. As you take this mental inventory you will notice your posture return to an appropriate interested look. In a nutshell, you will find yourself playing the part of the interested and intelligent student. This is called a "*mind set.*" It is just as important as the professional athlete who puts on a *game face*, except that you are putting on your thinking cap!

I KNOW EXACTLY WHAT YOU DIDN'T SAY!

In Chapter Five we learned about modeling: how everyone has their own individual style of stating their ideas and demonstrating behaviors. Communication is only 7% to 12% verbal; all the rest is non-verbal. When the awakened genius understands how a person uses non-verbal communication, he holds the key to knowing what is most important to that individual.

How can you use this in your education? Review Chapter 2, *Learning with Style,* and Chapter 5, *The Power of Modeling,* just before attending your class or lecture. If you notice the way a teacher uses his or her language patterns and body posture, you will begin to understand what is important to that teacher and what you will need to know for that particular instructor's tests.

FIVE KEYS TO PERFECTING YOUR LISTENING SKILLS

1. Get into **optimum learning state**. (Circle of Power)
2. Use the **three finger technique**.
3. Look for communication cues, such as pointing, using tonal anchors, underlining items on the chalk board.
4. Read your lecturer. Look for the obvious. Ask yourself questions such as: When might I be tested on this material? What information is the teacher stressing as important?
5. Take short notes that will help you to remember the highlighted information. Use a tape recorder if it is allowed and listen to it again in a relaxed setting with soothing music in the background.

TAKING TESTS WITH EASE

Geniuses learn to use time to their advantage. They are never wasteful. Even rest and sleep time can be used to your advantage. Each night before entering into sleep allow that part of your mind that is *other-than-conscious* to categorize and organize all the new thoughts and ideas you have learned through the day. This technique is very simple and will become automatic with practice.

1. **As soon as you get into bed, begin to notice the flow of air into the lungs and the flow of air from the lungs and let your body go loose and relaxed.** *Some people find that it is a much nicer experience if they relax each muscle group one at a time. Experiment to find which way works best for you.*

2. **Mentally start a process of imagining that you are taking your next test and all the answers are coming to you--indeed, you know every answer.** *Imagine what you learned on the first day of the semester all the way through the last day of school, refreshing your memory so that during the testing situation you can use the "sponge technique" (See Chapter 4, Incredible You!).*

3. **Continue the process of review until you feel comfortable with the information *(not that you consciously know it all, just that you know you know it all).*** *Then imagine what your family and friends will say to you when you receive 100% or 90% on the test.*

4. **Then move your thoughts to an activity of enjoyment for you.** *If you like sports, think of a sport you like to play and what it feels like when you are having fun. With those thoughts let yourself drift off to sleep.*

FIVE SIMPLE STEPS TO TEST MASTERY

The day of the test follow the steps below to optimize your mental state and receive the greatest help from your *other-than-conscious* mind.

1. Remember, your mind stores all questions with their answers, but, just like with a computer, the information must first be entered. **Be prepared** by attending all lectures and classes and by completing all homework assignments.
2. Get into your **optimum learning state**. (Circle of Power)
3. Use the **three finger technique**.
4. Before starting the exam, use your imagination to mentally review your notes and lectures.
5. Take three deep breaths, relax and, **one at a time**, answer all questions.

CHAPTER TWELVE
GOALS AND OUTCOMES

There is a big difference between ***goals*** and ***outcomes***. Goals are attainable but just out of reach. Outcomes are what you expect to have happen if you follow a blueprint for success.

The easiest way to distinguish between goals and outcomes is as follows: ***Outcomes*** for this process will be used for daily planning and weekly programming. ***Goals*** are to be used for monthly and yearly projections.

SELF-DISCOVERY: SUCCESS PROGRAMMING

First thing in the morning make a list of outcomes for the day -- list only those items that you need to get done today and that will fit with your schedule. (This list can also be written out the night before.)

One day a week make a separate list of outcomes for the week. Carry your outcome lists with you and keep them as visible and available as possible. If you ever find yourself feeling lost or uncertain as to your next move, take a look at your outcome list. It will put you right back on track.

Then create a list of goals with the understanding that each daily and weekly outcome will be completed.

- **Goals for the next month?**
- **Goals for three months?**
- **Goals for six months?**
- **Goals for one year?**
- **Goals for five years?**

After you have written out your goals on a piece of paper, post the list in a visible place where you will be able to read them at least once a day. This focuses your mental energy in the direction of your goals and helps to keep you on schedule. Never be afraid to revise your goal list. Life is ever-changing and so your goals should be also. If you become uncertain about a goal, list it at the bottom of the page under "Possibilities." Nothing should be ruled out as impossible. The genius knows there are no limits.

SELF DISCOVERY: LIVING YOUR DREAMS

Imagine for a moment that you are attending a funeral. You are sitting in the back of the room. There is someone for whom you hold a great deal of respect up at the podium speaking about the wonderful contributions this person has made to his or her community. As you listen you begin to realize a common thread of integrity that ran through this person's life. This was someone motivated to help others just as he or she helped him/her self. There was a true win/win attitude shining through all of his or her accomplishments. Slowly, as the speaker finishes up, you hear a brief recap of all that this person had achieved in one lifetime. Just at the end you realize that the person they have been talking about was you.

What would you like to have heard spoken from that podium about you?

Naturally, if it actually got to the point of the above scenario, you would no longer be in control of your life. Your life would be completely behind you. But, isn't it true that you only exist in this moment, **now**, anyway? Which is, by the way, the most powerful moment there is. Your past is behind you and your future is before you, and what you do today for those you love and the people you meet will be your greatest legacy. Being of service to humanity is the lasting sign of a genius. As you add to life, so life will add to you.

HOW IMPORTANT IS YOUR SELF-IMAGE?

Research has shown that the self-image is directly responsible for attitudes and helps to formulate beliefs. It has also been found that people with a poor self-image tend to be overweight and can lean toward usage of drugs or alcohol. The self-image can often seem somewhat elusive. You can't go to the corner market and buy a pint of self-esteem. Truly self-esteem exists in a gray area that no one can really define. But there is help. In my experience, an individual with low self-esteem is one who is operating in survival mode. Just like a wild animal, if one is *reacting* to life instead of *interacting* with life the stress will cloud better judgment and doubt will creep in. It would surely be nice to just run to the corner store each time you needed a boost of self-confidence or self-esteem, but, as of the writing of this book, that is not yet available. Until such time, we must trust in our minds to create these positive feelings for us.

One of the difficulties NASA encountered while developing the space program involved the astronauts' reaction time. Problems would arise faster than the astronaut was able to consciously respond. Because of this, NASA developed a training program that taught astronauts to think in *future time*. While on a mission, the astronauts were then in a state that allowed them to continuously forecast the future. These astronauts found that once they trained their minds to think in *future time*, they could detect problems before they occurred. Otherwise, it would be too late.

Similarly, if you become aware of your own launch programs (such as how you talk to yourself or picture yourself in the future), you will recognize how they directly relate to feelings of self-confidence and self-esteem.

Remember, many studies have shown that the mind and body cannot recognize the difference between real or imagined. So the secret to creating a powerful self-image is to create success in your mind and then feel it happen in your body. A proper affirmation will create a movie on the inner screen of the mind; one which is attainable by you. From there you will recognize the truth in the statement, *what the mind can conceive and believe the body will naturally achieve.*

TIPS TO BUILDING A POWERFUL SELF-IMAGE WITH AFFIRMATIONS:

An affirmation is the use of words or statements to elicit a positive response and outcome. You can use these words and statements every day to help you visualize your goals. By affirming what you want each day, you are doing something positive toward the attainment of your goals. It has been said that a picture is worth a thousand words. If this is so, one might wonder how a spoken word is going to change a self-image. The key is to use language patterns that create the pictures you want. By using positive affirmations you will become the "Steven Spielberg" of your mind. You are the producer and director of your inner theater; you are directing your mind to create what you want.

TIPS FOR WRITING AND USING AFFIRMATIONS

Always state what you want in the first person (*I am* or *I will*). State your affirmation in the positive and as if it has already occurred. *For examples of affirmations, and to increase your learning abilities and your self-image, look through the Self-Help Dialogues, choose from the list below, and use your creativity to make up your own.*

Write all affirmations in your own handwriting. Record all affirmations in your own voice. Read the affirmations aloud at least once a day so you can be reminded of their intrinsic value. Your mind will accept your voice much quicker than that of another person. Most people have spent the greater part of their life telling themselves what not to do and we all know that doesn't work! Your mind is starving for the loving tone of your voice reinforcing positive thoughts and patterns.

AFFIRMATIONS FOR LEARNING

The world is a safe place to learn.
It is exciting for me to learn.
I am at peace with educational material.

I am activating my learning potential
I am intelligent and can learn.
My mind is open and receptive to new ideas.

Intelligence, courage and self-worth are always present in me.
I have the courage to succeed.
I can make time to improve my life.

It is okay to study and enjoy life.
I can see myself getting an "A" on my exams.
I can let my mind wander and think positive thoughts.

I can understand lectures.
Taking tests is fun. I look forward to testing.
Everything I see, I remember.

Everything I hear, I remember.
Everything I experience, I remember.
My memory works in all situations.

My memory improves every day and in every way.
It is easy to relax and to learn.
My study habits improve daily.

I am taking steps toward accomplishing my goals.
I can do something today to improve my life.
I can plan for a bright and compelling future.

I can enjoy school.
My ability to concentrate improves daily.
I can take steps to improve my life -- it's safe to improve my life.

I'm comfortable with change.
I trust the world and accept all changes that are necessary.
Today is a great day to improve my mind.

I choose to lead a productive and positive life.
I can set an example for others through my positive actions.
It's perfectly okay for me to say "NO" to wrong actions or ideas.

It's perfectly okay for me to say "YES" to myself.
I'm free from past limitations.
I accept my mind and its full abilities.

Life is great! I can lead a successful, positive life.
I'm safe to succeed.
I'm flexible and can make friends easily.

Success is a state of mind.
I'm willing to believe in myself and bring about success--I'm free!
My reading abilities are constantly improving.

My ability to recall information is improving.
I can be myself.
Negative thoughts and influences no longer control me--I'm free!

I can accelerate in the classroom.
I stay focused on the task at hand.
I can see, hear, and feel myself improving every day and every way.

My mind is like a sponge--in stressful situations a fresh new idea is squeezed from it.

AFFIRMATIONS FOR LIVING

I am centered and calm and balanced.
The universe approves of me.
I trust my *other-than-conscious* mind.
All is well--I am in harmony with the universe and all of life.
It is safe for me to know and grow.
I am responsible only for myself and I rejoice in who I am.

I can handle all that I create.
I am clear in my communication with life.
I am free to enjoy life right now.
My communication is clear.
I accept what is good for me.
I let go of all expectations.

I am loved and I am safe.
I lovingly release others to their own lessons.
I lovingly care for myself.
I move with ease through life.
I have a right to be me.
I forgive the past.

I know who I am.
I touch others with love.
I accept life and I take it in easily.
All that is good for me is mine now.
My heart forgives and releases.

It is safe to love myself.
Inner peace is my goal.
I forgive everyone; I forgive myself.
I nourish myself.
I give myself the gift of forgiveness and we are both free.
I let life flow through me.
I am willing to live.
All is well.
I trust life to unfold before me in positive ways.
It is safe to love myself.

I willingly let go and allow sweetness to fill my life.
I am open and receptive to all that is good for me.

The universe loves me and supports me.
I claim my own power.
I lovingly create my own reality.
I open myself to joy and love which I give freely and receive freely.
I see myself as beautiful and lovable and appreciated.
I am proud to be me.

I choose to circulate the joys of life.
I am willing to nourish myself.
I am safe in the universe and all life loves me and supports me.
I grow beyond parental limitations and live for myself.
It is my turn now.
I release the past.

I cherish myself.
I am safe.
I am loved.
I love who I am.
I am grounded in my own power.
I am secure on all levels.

I deserve to enjoy life.
I ask for what I want and I accept with joy and pleasure.
I am the power and authority in my life.
I now release the past and claim what is good for me.
I bring my life into balance by loving myself.
I live in today and love who I am.

Most of the true geniuses, such as Nikola Tesla, had little concern for *fame* or *riches*. These are simply the natural by-products of doing what you love. If you don't know who Nikola Tesla was, don't feel alone. Very few people are familiar with Tesla's inventions. Yet, his discoveries have had an incredible impact on the development of society as we know it today. Tesla is best known as the person responsible for creating alternating current (AC), which he patented and later sold to George Westinghouse. He had a vision of the world where all people could share in free energy. Sadly, many of Tesla's inventions never made it because there was no profit to be made in free energy. While he was alive he never obtained the fame or wealth that he rightly deserved. Only now is he beginning to receive the recognition he merited for his many great inventions.

The Tesla Coil is still in use today and many of his other inventions, like Leonardo da Vinci's, will take years, perhaps even millennia, to come to fruition. How many other geniuses have never been given the recognition they rightly merit? We can only speculate. But like travelers who know where they are going and how fast they must drive to arrive safely, over four billion geniuses are now equipped to take the path of discovery. When each "genius in training" passes into the great *undiscovered country,* each will leave behind a legacy in his or her own way.

So like the best known genius of our time, Albert Einstein, who believed life was exciting and sleep was a waste, we could collectively start living our dreams; building a world of common everyday geniuses working together to leave a legacy of growth and development. Discovery of the human potential, the truly *undiscovered country,* is the three-pound universe that exists in each of our heads.

***"We don't inherit the earth from our ancestors,
We borrow it from our children."***

"The great thing in this world
is not so much where we are,
but in what direction
we are moving."

Oliver Wendell Holmes

CHAPTER THIRTEEN
A GENIUS IS PROACTIVE

It was one of those spring storms that came out of nowhere. The beach was chaos as tourists and local vendors scattered in every direction trying to find shelter from the storm. Monstrous waves crashed against the shoreline, they ripped claw-like into the sand, digging deep rifts where children had been building sand castles only minutes earlier. Mother Nature had proved her supremacy once again.

Just as quickly as the storm had appeared, it now vanished into a distant gray rumble. The beach was cleared of the human clutter, but a new kind of chaos took its place; it was breathtaking as the sunlight danced across a sparkling array of stranded starfish. There were thousands of them scattered everywhere. As the children returned to the beach, some of the boys began to throw the starfish back into the now calm sea. It gave them a sense of pride to know they were helping the marooned little creatures.

Just then an elderly man, Mr. Jones, approached. He walked with an odd swinging gait as he slowly dragged a partially lame foot across the sand. Dull gray hair stood out in tufts and a full yellowed beard surrounded his face. His left eye had an ancient scar that tugged the lid upward. He had a forbidding air about him and most children moved in another direction whenever he came around. He liked it that way.

The old man stopped his halting gait and watched the boys for a moment. "Stop that," he called, "Can't you see that what you're doing isn't going to make any difference ...none at all!? This beach goes on for hundreds of miles! It's loaded with stranded starfish just like these. What you're doing won't matter a bit, not one bit!"

The boys stood as if frozen and eyed the wintry old man, unsure of how to respond to his demands. The youngest of the boys, Jimmy, gazed thoughtfully at the starfish laying closest to his feet. He quietly picked up the starfish and cast it out into the ocean. He turned to Mr. Jones and calmly stated, "It mattered to that one."

HABITS OF THE GENIUS

What we will be learning about in this next section is how, by awakening your genius, you can anticipate what might occur around you and stimulate a more appropriate state of mind instantly and without putting forth much conscious effort. We all have "Mr. Jones," the pessimist, within us, but we have "Jimmy" there too. A genius takes control and dwells on the attitude of "Jimmy" and knows that even the smallest change is worth the effort. You will be transforming past negative tendencies into the habits of the genius.

We all have some little habits or quirks that we would like to change. Perhaps you would like to lose those few extra pounds, stop smoking, rid yourself of a chemical dependency or put an end to your procrastination. Plus, everyone can benefit from enhancing their individual talents and abilities. How about those study, reading, or public speaking skills? Maybe you just want to swing your golf club like a pro. This next pattern will show you how, by taking three minutes a day, you can train your brain to create experiences that flow in the direction of success; instead of waiting around for experiences to which it can react.

A *habit* is a way in which you normally respond without a conscious thought. One example would be automatically opening the refrigerator door when you enter into the kitchen. Have you ever found yourself standing with the refrigerator door open only to realize that you aren't hungry, but habit somehow led you to open the door and gaze inside?

A true genius understands that a habit is something done without thinking and that there are good habits and bad habits. The following process is designed to build genius habits by making you aware of your own habitual nature and transforming any negative behaviors into positive resourceful responses. If you are one of those fortunate people who has no unwanted reactions like anger, fear or frustration, simply change where it says "negative" to "positive" reaction. The process will work either way.

SELF-DISCOVERY: QUANTUM PROCESSING

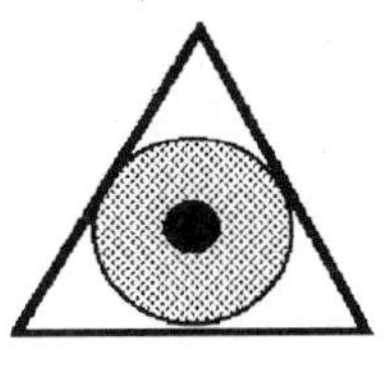

Throughout this process you will be asked to use a "focal point." This is the illustration you see on this page and on the following pages of the Quantum Processing section. Simply read through the instructions and then use the "focal point" as directed.

You will also be asked to create what is known as a "break state." Because the mind is sequential in its learning, it needs a break in the sequence so all information can be processed in the most complete and appropriate manner. When you are asked to "Break State," simply lift your eyes away from the book and notice something specific in the room around you. Perhaps you will look at a lamp, a chair, a piece of art, or gaze out the window at a tree. Simply divert your attention for a moment and then return to the focal point.

Begin by imagining what your life can be like if you spend just a few minutes each day to set up success. *(Remember, your mind works best with specifics.)*

1. Think of one specific habit or tendency which you would like to change. *Remember, the genius realizes that he can only change that of which he has control; which equates to self control or control over the senses.*

2. Think of what it's like the moment ***before*** the emotion starts. *Become aware of the reaction of your body. Whatever it is, be there in the moment. See what you would be seeing, hear the sounds that would be around you and feel as if you are there having the experience some time in your past. And, when you have it, hold on to the feeling for a moment.*

3. BREAK STATE *(Divert your attention by looking at something in the room. Move your body a bit to relieve any unwanted feelings.)*

4. Focus your attention on the **FOCAL POINT**

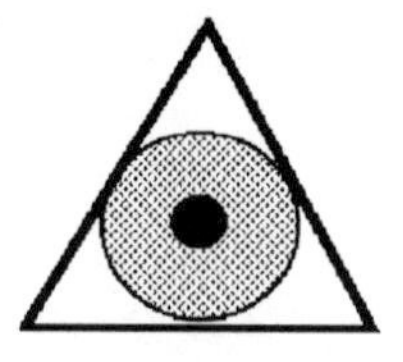

5. As you think in your imagination about that focal point, I want you to create in the center of it a very small picture of you. You really can't see it yet. Imagine it deep in the middle of that small circle. From there I would like you to put into this image all the qualities you would put into the ideal you ...Your looks? ...Facial features? ...Your weight? ... The way you dress? ...The way you smile? Be as creative as you can. Leave the image in the middle of that focal point ...What would that image be saying to itself? This image of you creates goals and then attains them easily and without effort ...It has all the positive changes that you would ever want, need or desire ...And the best part is they all work without effort ...they simply work where you need them the most in your future.
6. As your mind continues to build into this image all the skills, all the abilities, and all the resources that you would ever need, slowly imagine that image sliding out of the focal point and moving toward you. It becomes a bit bigger now, a bit brighter, until it is right there in front of you. Then, automatically, when it is there, just close your eyes and imagine that this new you fills your body ...every cell, every system and every organ ...totally and completely. And when you are completely full of that feeling, just open your eyes,
7. BREAK STATE *(Return to the room and look around.)*
8. Now, as vividly as possible, bring back the moment that occurs before the unwanted emotion or habit. Once you have this feeling, imagine that even while you are experiencing its feelings, sights and sounds, you are being drawn to that focal point ...you know all that it holds for you.

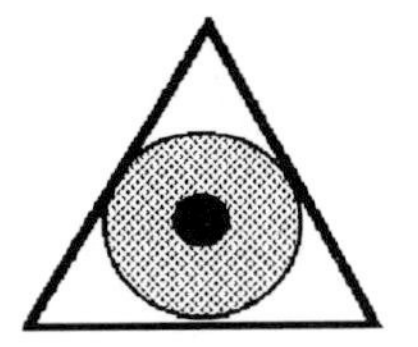

9. Because you have read through the process before, you can now, without a conscious thought, bring the image from your focal point and slide it closer and closer to you until the image once again fills your mind ...it is now the image of you successfully handling the situation. Whatever the situation, you remember that your internal genius can handle it far better than you can. At that moment you can let go of conscious control, and you act as if the situation is complete and you allowed the best behavior to flow to the surface. I'm not sure exactly how you knew that in your focal point was the new resourceful you, but somehow you knew. Instantly and automatically the new you now comes into view right in front of you.
10. You can now close your eyes and allow the positive feelings and a new level of understanding to occur so that you can imagine acting and responding differently to the old stimuli. When you have those positive feelings in and around you, begin to think where in the future you would want such a behavior if it were to occur without your conscious mind ...but it just happened. Maybe when you walk through a doorway ...maybe when you turn a light switch on ...and, of course, as you need it and whenever you need it the most.

As with any strategy for success, this one will need to be practiced. A genius understands the benefits of mental practice. There are only so many minutes in a day, but when you start to use your imagination there are no limits to what can happen. Remember that five minutes of visualization is worth two hours of physical practice. Awakened geniuses know how to use time to their advantage. Remember your brain is always striving to prove you right. ***You will get what you rehearse!***

Take the time each night before entering into sleep to rehearse the way in which you will respond to people as well as your environment and, just like the Boy Scouts, always be prepared. The ability to succeed in life lies in flexibility and adaptability; the ability to manage frustration and create new responses.

"You can have anything you want if you want it desperately enough. You must want it with an inner exuberance that erupts through the skin and joins the energy that created the world."

Sheila Graham

CHAPTER FOURTEEN
A GLIMPSE INTO THE FUTURE

I invite you to come with me on a journey into the future--to a world of the awakened genius. It is the dawn of a new millennium where our school systems have advanced to integrating computers, technology and quantum learning methods to awaken the genius in each child. The year is 1999 and the first full spectrum Holographic Learning Machine is in use. Could learning machines really be available that soon? My answer is an emphatic YES, because learning technology is here today.

WHERE WE STARTED

It all started back in 1924 with Hans Berger, a German psychiatrist, who labeled the "Alpha" frequency and published "wavy-line" pictures showing electrical activity of a human brain. This led to the development of a new science of Electroencephalography (EEG).

The field continued to expand, and in 1949 the Tuposcope was invented, which gave scientists and physicians the ability to track the Beta, Alpha, Theta and Delta brain wave patterns. In the 1950's and 60's with this device and the use of Yoga, Zen and Buddhist meditations, it was found by M.A. Wagner of the University of Michigan School of Medicine that the electrical activity of the brain changed when an individual moved into the meditative state.

Then came the quantum leap in brain wave education and research. In 1980, Marshal Gilula, MD, Life Energies Research Institute of Coconut Grove, Florida, researched light and sound equipment and published a paper entitled "Multiple Afferent Sensory Stimulation" (MASS). His research revealed that 80% of his patients experienced deep muscle relaxation and deep states of mind-body relaxation when they used light and sound equipment.

This light and sound equipment uses passive audio-visual stimulus to produce a relaxed pattern for the brain to follow. A gentle pulsating light travels through closed eyes and a sound wave enters the auditory canal and the frequencies harmonize. The pattern gently guides the brain into the Theta state of consciousness. By helping to synchronize and focus the activities of both the left (logical) and right (creative) hemispheres of the brain, the user is placed in an ideal learning state.

Where We Are

Today students and business people are using light and sound technology to improve their performance in all areas of life. In 1987, I was fortunate to have been involved with the developers of a personal light and sound device known as the MC² (square). It was the first portable light and sound device to be marketed to the general public. The founders, Linnea Reid and Larry Gillen, conducted extensive research with people who were using the equipment on a regular basis and found it beneficial for nearly all applications, from basic stress reduction to super learning.

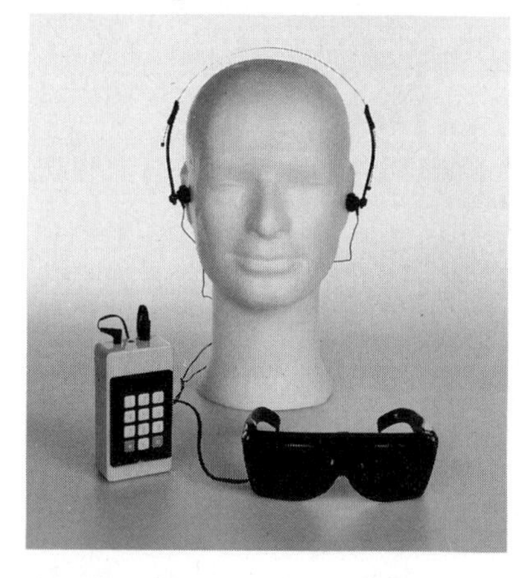

We have continued researching the benefits of light and sound technology in our own clinic, and what we have witnessed holds great promise for the future. Nearly one hundred percent of those students we have seen in our clinic as participants in the Awaken the Genius Project have found success. Each of them experienced the methods outlined in this book along with regular use of light and sound technology. We have seen students go from C's and D's to A's and B's. Children who once hated school now look forward to their day-to-day involvement in classroom activities.

Destiny is not a matter of chance; it is a matter of choice.
It is not a thing to be waited for; it is a thing to be achieved.
William Jennings Bryant

WHERE WE ARE HEADED

Where can we go with this wonderful machine and the many others like it? Medical doctors who work with the brain already know that if they stimulate certain areas of the brain, the person will have a full representation of a specific memory. In other words, the individual is sensing their total presence in the experience; it is holographic. They can see, hear, feel, smell and taste everything as if actually present at the time the event happened. Imagine where this discovery and its advancement could take us--when a future genius discovers a way to train the mind to access those holographic memories instantly. This single advancement would render most of our current school system obsolete. About 75% or more of each school year is geared toward review and testing. Once a student is trained in super memory processes, that time could be spent on research and creative discoveries.

Lets take a ride into a typical school in the year 2011. This is the year the holographic learning module is in full use. Instead of the student reading about outer space by tediously studying a book, he or she will actually be taken there by sitting in a comfortable reclining chair and allowing the universe to unfold.

Once seated, a large helmet will come down and block out all outside noises. The student, let's say her name is Mary, then begins the lesson for the day. The class is learning about the solar system, and Mary's first stop is the moon. It's a wondrous journey. Mary feels as if she is there, on the moon; all of her senses are present in the experience. Mary is being shown the space labs where they are conducting ongoing research for super learning in the non-gravitational state. Mary now learns all the geographical facts about the moon. She will have 100% recall as the facts are coded chemically into her core memory through the Holographic Imager. Mary will never be required to review what she has learned here. It has all been stored as permanent fact within her memory banks.

What students of the Twentieth Century learned about math and English in their elementary school years is being mastered during the sleep state while still small children. The planet as a whole began to speak one language in 1999 as mandated by the World Council of Nations. Education is elevated to the state where each individual is more involved with creativity and making a better planet than greed and self-centeredness. It has become unnecessary for any person to work more than 30 hours in any given week. Through unprecedented technical advances the destruction of the planet has been reversed, and people travel in vehicles that use the gravitation waves of the planet's surface. Energy is free and non-polluting with the use of the technology developed in the 1930's by Nikola Tesla, so free travel to all areas of the planet is available to everyone. There is no longer a need for paper; everything is stored electronically. Think of it, no paper mills polluting our air, no clogged land fills, and no clutter! The average IQ is well over 180, and as a group the human race is continuing to discover that the only limits are the ones *we* place into consciousness.

"Eighty percent of success is showing up."
Woody Allen

Is This Future Possible?

I plan on being alive to witness many of these changes as the geniuses of the future are born and come of age. Perhaps one of you reading this book and awakening your genius will be that person who invents the circuit that makes it possible to train the brain to be whole in its thinking. Or perhaps it will happen all on its own. Maybe you will be the "hundredth monkey" -- the genius mind that inspires a collective change on planet Earth.

It's probably hard for you to imagine a simpler time, but think about your grandparents' lives. Most of them grew up without televisions, VCR's or computers. They didn't even have video games! Yet those who will read these pages probably take computers for granted. But did you know that in 1970 no one ever heard of a personal computer?

When I was growing up in the 60's, I remember thinking that the world couldn't possibly get any more modern than it already was. What was there left to invent? But the computers of that time took up complete buildings and couldn't do nearly the processing that a typical personal system can do today. What about CD's and the storage of information? It is now possible to access information at such an alarming rate that truly anything is conceivable.

Are you willing to believe in a future where peace and the pursuit of collective happiness are the founding principles? We are on a journey -- it will not end with us, just as it did not begin with us. What part of your genius will you leave behind for future generations to use? It is my hope and prayer that you will choose to be a herald of the new generation; that with your awakened genius you will strive to surpass those greedy and mistrustful individuals who are currently leaving a legacy of pollution and social injustice in the pursuit of their own desires.

True geniuses work to benefit all of humankind, knowing that as they help those around them, they are naturally helping themselves. I commend you for the commitment you have made to awakening your full potential, to building a planet of harmony, and to developing and living your dreams.

"Have you heard Natalie Cole, daughter of the late Nat King Cole, sing with her deceased father? ... Beautiful, wasn't it? ... That was impossible until a new tool of technology allowed her to separate her father's voice from the rest of the sounds on the recording. A new tool allowed her to achieve what was to her an impossible, ideal goal. ... She used a new tool to change the rules... New tools, therefore, allow you to achieve new goals, redefine your role, and change the rules of the game."

Daniel Burris
TechnoTrends

Life is like a dogsled team.
If you ain't the lead dog,
the scenery never changes.

Lewis Grizzard

CHAPTER FIFTEEN
INSTANT GENIUS

Now that you have come to the end of the text, you might be thinking to yourself, "Am I a genius?" My answer is a most emphatic YES! You could still have problems popping the top off the IQ charts. But now you know that IQ is not an accurate test of genius. Now you are your own judge. For you, the real test will be in putting together a life of fulfilled dreams or, as most people call them, goals. I have heard it said that people don't plan to fail, they fail to plan! If you have not yet put together a game plan for your life, now is the time. Go ahead, make your goals real and tangible by putting them down on paper (See Chapter 12). Dare to truly dream your dreams. Remember what the good book says, "Faith without works is dead." This is precisely what Edison meant when he said genius is 1% inspiration and 99% perspiration. It is not always the swiftest person who wins the race. Rather, it is the individual who *knows* he can do it--and then goes out and overcomes all obstacles to *make* it happen.

"I DON'T HAVE TIME!"

The reality is you don't have time not to! Most people spend more time planning a weekend get away than planning their life. The two most valuable times of your day usually come and go without much notice. One such moment comes right before falling asleep at night and the other is the time you spend awakening in the morning. These are times when your mind is already in a perfect state for creating what you want. How can this time be used to your advantage? These priceless moments are perfect for success programming (See Chapter 12). Right before entering into sleep simply read your goals and any affirmations you may like and then fall asleep with one of your Self-Help Dialogues. In the morning, before polluting your mind with the negativity of the newspaper or T.V. talk shows, take time to read a few pages from a motivating or inspirational book or listen to tapes by the kind of people you want to model (See Chapter Five). Time is your most valuable asset; just ask that senior citizen who regularly complains about how much time he or she wasted in meaningless pursuits.

What about those idle moments when you are standing in line or waiting for the stop light to turn green? This is a perfect time for you to relieve the stress of the day by focusing on your breathing. Focused breathing is a technique of breathing in deeply and then breathing out completely. The "in" breath and the "out" breath should be of the same duration. People find it helpful to focus on the sound "***so***" while breathing in and the sound of "***hmm***" while breathing out. You will be amazed by the results you get by practicing the art of focused breathing during a time that would have been otherwise wasted. The main block to genius is stress. By practicing relaxation, you will be removing the blocks that may have prevented success in the past.

As I stood before my father, boldly pronouncing that one day I would be captain of the football team, I desperately wanted to be that famed athlete right then and there. Fortunately, my father took the time to teach me the benefits of patience and perseverance. I was able to discover how much fun the journey could be. Success is not a destination; it is a state of mind. The most important step you can take right this moment is to commit to spending time on you, time to build your successful state of mind. From there you will enter into each new encounter with the attitude that there is a greater reason for your presence than what you consciously know about.

With the decision to improve your life, you can read over this text once again and do each exercise with a renewed sense of excitement in self-discovery. Believe it or not, you are your greatest investment and most prized asset.

"I learn the way a monkey learns -- by watching its parents."

Queen Elizabeth II

THE 21-DAY PLAN

I recommend that you start with a 21-day plan using the tools set out in this text. For your first 21-day plan create realistic goals that can prove the power of your mind to you. Studies have shown that when you practice a new behavior for 21 days you have built the neuro-channels that make the behavior permanent. Remember, your mind will create what you are rehearsing. If that's not what you are intending, it's time for a change. All you have to lose is a little ignorance!

At the end of 21 days review your plan. Make note of the successes and anything upon which you need more emphasis. Always remember, there are no failures in life, there is only feedback. Now you are ready for the next 21 day plan...and then the next. Each 21-day plan creates new and more exciting self discoveries. Just watch yourself grow!

As an awakened genius you have planted the seeds to grow a garden of greatness. The techniques in this book are designed to be used on an on-going basis to keep your garden free from weeds and pests. It is time for you to invest in yourself. If you haven't already done so, go back and make your Self-Help Dialogues, then listen to yourself regularly. Your mind has been conditioned by the situations and circumstances of your life. Now you are ready to change your perception of the past and unleash your bright and exciting future. Good luck, Genius, and remember: life gives in the exact proportion that you are willing to accept.

"Realize the Heaven within you and all at once all desires are fulfilled, all misery and suffering is put an end to. Feel yourselves above the body and its environments, above the mind and its motives, above the thoughts of success or fear. The great cause of suffering in the world is that people do not look within, they rely on outside forces."

Baba

INDEX

ALTERNATIVE RESOURCES

Positive Changes Centers: Each of the Positive Changes locations operates the Light & Sound technology for learning as described in this book. Please call the appropriate office to schedule a free introductory session. Experience the excitement of advanced learning technology. To receive a free booklet outlining our programs, please call the office near you.

NATIONAL HEADQUARTERS

Positive Changes Hypnosis
Patrick & Cynthia Porter
Managing Directors
Contact: Karen Schipula
5045 Admiral Wright Road
Virginia Beach, Virginia
www.pchypnosis.com
(757) 499-5097

Positive Changes Hypnosis
Newport News, Virginia
(757) 249-3082

Positive Changes Hypnosis
Chesapeake, Virginia
(757) 549-9208

AUSTRALIA
Dr. Mumtaz Vishal
Drysdale, Australia (03) 5251-2200

CALIFORNIA
Arun Bhojani
Fremont, CA (510) 797-7588
bhojani.arun@ssd.loral.com

Positive Changes Hypnosis
Fountain Valley, CA (714) 965-6999
bobharris@earthlink.net

Positive Changes Hypnosis
Santa Barbara, CA (805) 569-1202
hypnosis@silcom.com

Positive Changes Hypnosis
Sherman Oaks, CA (818) 986-9208

PCH of Escondido
Escondido, CA (760) 233-8800
hypnopc@aol.com

Positive Changes Hypnosis
Laguna Hills, CA (949) 951-9357
hypmcnatt@aol.com

PCH of San Jose
San Jose, CA (408) 247-5828
dhr5@aol.com

PCH of North County
Encinitas, CA (760) 635-2790

PCH of San Diego
San Diego, CA (619) 571-4722
hypno@cts.com

PCH of Santa Rosa
Santa Rosa, CA (707) 575- 4509
lmstewart@juno.com

PCH of Beverly Hills
Beverly Hills, CA (310) 657-6868

COLORADO
Maggie Connor
Boulder, CO (303) 998-0283
nrgmaggie@aol.com

CONNECTICUT
PCH of Milford
Milford, CT (203) 876-3393
miklj@juno.com

PCH of Ridgefield
Ridgefield, CT (203) 438-0255
bodymind@geocities.com

FLORIDA
PCH of Tampa Bay
Largo, FL (727) 519-0707
hypnolargo@aol.com

The Hypnosis Center
Fort Lauderdale, FL (954) 565-3517
smokeles@gate.net

PCH of San Fransico
vb1492@hotmail.com

Positive Changes Hypnosis
Altamonte Springs, FL
(407) 262-9100
amdcl@aol.com

Feel Well Hypnosis Center
Tallahassee, FL (850) 671-4555
Smokestop@aol.com

ILLINOIS
PCH of Chicago
Schaumburg, IL (847) 885-0815
pfaris@megsinet.net

Succesful Living Systems
Champaign, IL (217) 351-9630
thunder2@pdnt.com

PCH of Macomb
Macomb, IL (309) 837-4205
srunkle@macomb.com

PCH of Moline
Moline, IL (309) 793-4721

INDIANA
PCH of Fort Wayne
Fort Wayne, IN (219) 744-6869
csimon@fwi.com

IOWA
PCH of Sioux City
Sioux City, IA (712) 277-8634
HYPNO1@aol.com

MICHIGAN
Positive Changes Hypnosis at North Oakland Hypnosis Center
Oxford, MI (248) 628-3242

MINNESOTA
Southlake Hypnosis
Edina, MN (612) 832-5001

NEVADA
PCH of Las Vegas
Las Vegas, NV (702) 248-9030
VegasPCH@aolncom

NEW MEXICO
PCH of Albuquerque
Albuquerque, NM (505) 292-2237

NEW YORK
Dynamic Changes Hypnosis
New York, NY (212) 684-7608
Dchypnosis@aol.com

PCH & Counseling, Inc.
Rochester, NY (716) 865-8220
teall@aol.com

PCH of Buffalo
Buffalo, NY (716) 833-7272
byggtime@aol.com

PCH of Long Island
Ronkonkoma, NY (516) 580-2464
positivechanges@erols.com

PCH of Queens
Queens, NY (516) 371-3745
Slamoll@aol.com

OHIO
PCH of Dayton
Kettering, OH (937) 298-4939
jdelph@donet.com

PCH of Beavercreek
Beavercreek, OH (937) 429-1900
jdelph@donet.com

PCH of Toledo
Sylvania, OH (419) 882-8543
(419) 824-3176 - Fax
gryphon6@ix.netcom.com

PCH of Akron/Canton Ohio
Akron, OH 45409
(330) 645-7707

PCH of Cincinnati
Westchester, OH 45241
(937) 497-9181
novitek@bright.net

PENNSYLVANIA
PCH of Bethleham
Bethleham, PA (610) 691-5541
mindwrkr@bellatlantic.net

TEXAS
PCH of Austin
Austin, TX (512) 977-8700
dave@accessnlp.com

PCH of NW Houston
Houston, TX (281) 463-0647
benjh@texaco.com

PCH of NE Houston
Houston, TX (281) 447-5222
hgvetter@iamerica.net

PCH of SW Houston
Houston, TX 77057
(713) 977-4434

PCH of Irving
Irving, TX (972) 570-7747
(972) 438-6324 – Fax
pchdaltx@gateway.net

PCH of San Antonio
San Antonio, TX (210) 495-9933

UTAH
PCH of Salt Lake City
Salt Lake City, UT 84109
(801) 466-4044
hypnosis@slkc.uswest.net

PCH of Utah
Salt Lake City, UT (801) 569-9390
1mcrossi@burgoyne.com

VERMONT
PCH of Vermont
Montpelier, VT (802) 229-1365
padilla@plainfield.bypass.com

VIRGINIA
(See National Headquarters)

Positive Changes Institute
Fairfax, VA (703) 352-5042

PCH of Lynchburg/Roanoke, VA
Forrest, VA (804) 525-6230
Kirtlep.com@lynchburg.net

WASHINGTON STATE
PCH of Lake Stevens
Lake Stevens, WA (425) 377-9424
alldridger@aol.com

Positive Changes Hypnosis
Longview, WA (360) 577-1909
bryjames@cetnet.net

WISCONSIN
PCH of Green Bay
Green Bay, WI (920) 465-1277
mweist7124@aol.com

CANADA
Positive Changes Hypnosis
Abbotsford, British Columbia
(604) 826-7100
wendybeasley@yahoo.com

PCH of Canada
Raja, Satyen
Brampton, Ontario
CANADA L6W-3J7
(905) 454-7400
pchcanada@mailexcite.com

PCH of Calgary
Calgary, Alberta
(403) 209-0556
positivechanges@cadvision.com

Positive Changes Hypnosis
Toronto, Ontario
(416) 345-9885
earthangels@attcanada.net

Positive Changes Hypnosis
Oakville, Ontario
(905) 337-3700

EUROPE
Brown, Jamie
20 DeBeauvior Square
London, England N14LD

Goodey, Will
Englands Residence
25 Englands Lane RESIDENCE
London, England NW3 4 XJ

To receive a copy of our free publication
Positive Changes Hypnosis Today
and a complete list of Affiliates
Please call 800-880-0436.
www.pchypnosis.com

If you have a product or service that you would like to have listed as a resource, please send your proposals to Awaken the Genius Foundation, c\o 309 Aragona Blvd. PMB 102-712, Virginia Beach, Virginia state, Postal Zone/Code [23462].

Write for a free catalog of upcoming workshops & materials.

About the Author

Patrick Porter lives with his wife, Cynthia, and two children, Cheree' and Alex, in Virginia Beach, Virginia. Patrick travels the world lecturing and teaching the skills found in his books and tapes. He is available for corporate training and speaking engagements. Currently he is working with his wife and artist Dean Sylvia in developing childrens books that teach positive living and human values.

About the Cartoonist

Dr. Porter met Dean Sylvia while developing Awaken the Genius with the Arizona Health Council. Dean has since developed into one of the top aspiring cartoonists in the country. Dean Sylvia lives in Phoenix, Arizona where he is currently working with Dr. Porter and several authors in the development of childrens books for emerging geniuses. Dean has also developed "Just for Kicks!" a comic book series based on his years in martial arts. If you are interested in using Dean's talent for your project, he can be contacted at 602-780-3858.

About the Cover Artist:

Sam Davis Johnson is an illustrator from Lexington, Kentucky. The cover for Awaken the Genius was created with airbrush, colored pencil, pen & ink and brush techniques. Mr. Johnson's studio, ABS Airbrush Illustration & Design is engaged in artwork for the advertising and publishing industry. For more information 606-273-7075.

"YOU CAN HELP YOUR FRIENDS AWAKEN THEIR GENIUS."

Positive Changes and PureLight Publishing would like to provide you with an opportunity to transform our world. It has been said that we are only six people away from knowing everyone on the planet! Wow, that's exciting! If you would like to share Dr. Porter's concepts with your friends or family members, it's simple. Just send us their names and addresses and we will send them a free tape—a $9.98 value.

Yes! I want to help change the world into the type of place where genius is the norm. Please send the following people the

"Six Secrets of G.E.N.I.U.S" **Free** Tape.

(Please enclose $3.00 for the each set sent.)

Genius #1
Name:____________________________
Address: __________________________
City:_____________ St: ____ Zip:______
Phone: (   ) ____-_______

Genius #2
Name:____________________________
Address: __________________________
City:_____________ St: ____ Zip:______
Phone: (   ) ____-_______

Genius #3
Name:____________________________
Address: __________________________
City:_____________ St: ____ Zip:______
Phone: (   ) ____-_______

ORDER FORM

☎ Telephone orders: **800-880-0436 or 804-631-2928**

✉ Postal orders: Awaken the Genius Foundation, c\o 309 Aragona Blvd., Ste. 102-712, Virginia Beach, Virginia state, PZ (23462)

Please send the following products. I understand that I may return any products for a full refund, for any reason, no questions asked.

____

____

❑ Please send me the free Awaken the Genius tape. I have enclosed $3 shipping and handling.

❑ Please add my name to the Genius' Network so that I may receive more information on how to awaken my genius.

Name: ______________________________________
Address: ____________________________________
City: ____________________ State: _______ Zip: _____-____
Phone: ( ) ____-______

Sales tax:
Please add 6% sales tax for books shipped to Virginia addresses.

Shipping:
Book Rate: $3.00 for the first book and 75 cents for each additional book.
(Surface shipping may take three to four weeks.)
Air Mail: $4.50 per book.

Payment:
❑Check
❑Credit card:❑ Visa, ❑MasterCard, ❑Amex, ❑Discover
Card number: ______________________________
Name on card: ______________________ Exp. date: __/__
Signature : ________________________

Call and order now!